£1-50

A Tangle of Islands

A Tangle of Islands

L. R. HIGGINS

ROBERT HALE • LONDON

© L. R. Higgins *1971*
First published in Great Britain July 1971
Reprinted February 1972

ISBN 0 7091 2189 X

Robert Hale & Company
63 Old Brompton Road
London S.W. 7

PRINTED IN GREAT BRITAIN
BY COMPTON PRINTING LTD.
LONDON AND AYLESBURY

For

The Girl, Amanda Bridget,
the Boy, Charles Adrian,
who shared these adventures:
and Nicholas Patrick, who stayed
at home and drew maps.

Contents

———————————✠———————————

Illustrations

---·⚓·---

9

facing page

*Photographs on page 64 and crow's nest page 144 by C. A. Higgins;
all other photographs taken by the author*

Come! share my secret place,
a green jewel of an island
in a peacock sea;
carelessly spangled with lovely flowers,
and where only seabirds dwell.
You, too, can share my camp,
on a ledge, just wide enough to take a tent,
or a wee stone cot,
and a crackling fire of salted wood.
Forget the world of men, awhile,
and live with me through summer's days,
alone with the wild seafowl.

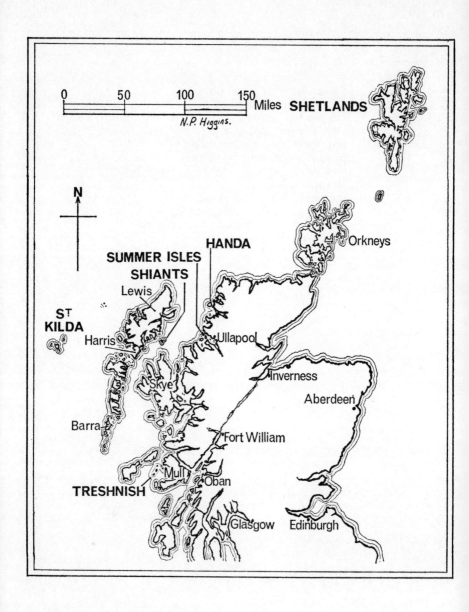

0   50   100   150 Miles   SHETLANDS

N.P. Higgins.

N

HANDA

SUMMER ISLES

SHIANTS

Lewis

Orkneys

Sᵀ
KILDA

Harris

Ullapool

Skye

Inverness

Aberdeen

Barra

Fort William

TRESHNISH

Mull

Oban

Glasgow   Edinburgh

ONE

The Land of the Simmer Dim

••✿••

Travel north far beyond the Pennines, farther north than the Scottish Highlands; leave the stormy Pentland Firth behind and you will come to the land of the Simmer Dim, a large group of islands straggling over 70 miles of seaway; as near to Norway as to Aberdeen; reaching past latitude 60 degrees to draw level with Greenland's icy tip, Cape Farewell. Here are green brown hills, scarred with the dark of peat hags, and splashed with the fleecy grey of diminutive sheep. Look east, look west, and there is the sea.

These are the Shetland Isles, more than 100 of them, about twenty inhabited; and this is as far north as you may go and remain in Britain. The Gulf Stream ensures an equable climate. Mild winters with long nights are enlivened by spectacular displays of the Northern Lights, the more bright because the air is crystal clear. The cool summers have remarkably short nights in midsummer; nights that are never totally dark; this the Shetlanders call the Simmer Dim.

We had no intention of going to Shetland, none at all. For three years we had explored the broads and rivers of Norfolk and Suffolk and were intending to go to Ireland and drift up and down the Shannon. To save time we would fly; and then I totted up the cost—it was alarming.

The Girl came home for the Christmas holidays full of enthusiasm and excitement about a new adventure; she and several of her college chums were going to Norway in the summer. They never went; they too totted up the cost.

Norway! It set me thinking. Not the towns and cities but the

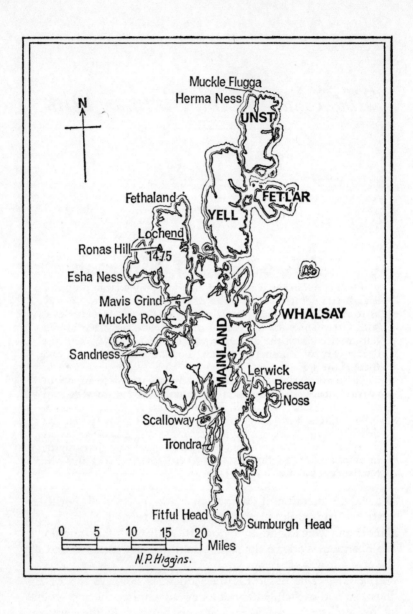

N

Muckle Flugga
Herma Ness
UNST

FETLAR
Fethaland
YELL
Lochend
Ronas Hill
1475
Esha Ness
Mavis Grind
Muckle Roe
WHALSAY

Sandness

MAINLAND

Lerwick
Bressay
Noss

Scalloway
Trondra

Fitful Head
Sumburgh Head

0   5   10   15   20
                        Miles
N.P.Higgins.

fishing villages and the fiords. Norway? Instead we went to
Shetland and never regretted the choice; the cost was half as
much again as the Shannon trip would have been. The next year
we went again. Someday I shall go back once more.

"Shetland usually is thought of as being rather remote a place
that very little is known about", states a booklet published by
the Shetland Tourist Association. How true. Cook's man stared
at me blankly. British Railways gasped incredulously: "Where?"
A letter to the Scottish Tourist Board—surely they knew?—
remained unanswered; three, four, five weeks passed, and then
came a wad of information from the Tourist Information Centre,
Lerwick, Shetland, to whom the Scottish folk must have passed
my enquiry. In the meantime, by rummaging around public
libraries, we began to paint a mind's-eye picture of the Shetland
Isles.

We started with our scant knowledge: Shetland is up north;
diminutive ponies came from there. That was the total. Piece by
piece we added to it: the biggest island is called Mainland and is
bigger than all the others put together; there are no trees and
hedges; with notable exceptions the hills are low, no more than
500 feet; these hills are covered with peat which is dug out and
burnt as fuel; the original inhabitants were Picts, but they
disappeared and the Vikings took possession; the islands belonged
to the kingdom of Norway for many centuries but a stony-broke
king pawned them to a Scottish king to cover a debt and never
redeemed them; the Shetlanders never liked the Scots any more
than the Irish like the English, and for similar reasons; the
Shetlanders are fishermen and crofters, and crofts are small-scale
farms carrying a few sheep and cattle—and few crops because the
soil is not rich; the women folk knit garments from Shetland
wool, and make fine lace; the population (about 18,000) slowly
declines and a third of the total is concentrated in Lerwick the
capital; many beautiful and rare wild flowers grow on the moors
and hills; the bird population is terrific; ships sail regularly
throughout the week from and to Aberdeen; Sundays excepted,
there is a daily air service from Aberdeen.

From total ignorance we emerged all-knowing. Back to Cook's.
Book three return tickets London–Aberdeen; northward overnight, on the Aberdonian, out of Kings Cross, arrive Aberdeen,

with luck at nine; never mind if it's late, we have six hours to spare. Book three return tickets, B.E.A. Aberdeen–Shetland; out Aberdeen 3 p.m.; in Sumburgh, 4 p.m. So simple. Shetland ". . . rather remote". Not any more; twenty hours from West Kent. Less if you travel by air all the way.

Further rummagings, on station bookstalls, had brought to light a book containing addresses and advertisements of houses, cottages and caravans to let. Tucked away on the back pages were two from Shetland, croft cottages, to be let to holiday makers, terms moderate. They were; the one we took cost us £3 a week.

Amongst the information sent by the Shetland Tourist Association was a very complete list of accommodation—hotels, boarding houses, cottages—available in all the islands, and a guide to the current charges and rents.

Within a couple of months, and before winter was out, we had all the arrangements complete; nothing to do, just wait patiently for the next six months to pass. Ideally, for my purpose, this was too long, but schools and colleges do not arrange holiday times to coincide with the breeding cycle of seabirds. They should do.

Economy is the Girl's watch word; on her advice I didn't book sleepers for the overnight rail journey. She counselled that it was quite comfy sleeping in a corner seat. I had to agree that this was true; I have done so much of it over the years. And, I was reminded, we want to see places we pass in the night. At York the pair of them, she and the Boy, were asleep; I nudged them; they opened sleepy eyes on the window, and promptly drooped their eyelids. Darlington came by; I nudged them again. The response was even weaker. They ought to see Newcastle, I mused; all I got was a bellicose look. Dawn and Edinburgh looked into our window simultaneously. This they must not miss. Edinburgh? So what! The Forth Bridge, I explained, we'll be going over soon. Five seconds was all they could spare. I gave up. As we rumbled into Dundee the pair of them returned to the world, of their own accord, looking very wan, like shipwrecked sailors; and both insisted they hadn't slept a wink.

Breakfast and an early lunch mended them. We flew to Shetland from Aberdeen in the afternoon, and in next-to-no-time were zooming over Sumburgh Head, walking off the pocket-handkerchief-sized airfield, and there was our car awaiting us.

This we hired from a Lerwick motor company so that we could travel the length and breadth of Mainland island. We did the length almost in one go, because our cottage at Lochend was in the northern part of Mainland, 65 miles from Sumburgh and almost at the end of the road.

We lingered over those miles for three hours; looking ahead, to left, to right, captivated by the Shetland scene. The plans we made as we loitered our way northward! We must see everything; we did too, but we had to come again. The ever changing scene held us spellbound; every corner and bend—and they were countless—revealed a new delight; there was the softness of the long narrow peninsula from Sumburgh Head to Lerwick; the wildness of the empty moors as we climbed the hills beyond. Then the scene changed, became more rugged; the sea was never far from sight, long inlets driving deep into the coast. We were quite unprepared for Mavis Grind; there Mainland is all but cut in two, narrowing to no more than 110 yards, this narrow strip keeping the North Sea and Atlantic Ocean apart. The road, cut through rock, hangs shelf-like by the North Sea on the one side, skirts the Atlantic on the other, climbs quickly and twists through hewn rock then descends rapidly to wind its way across wild and rocky moorland patchworked by lochs. The road, well surfaced and wide enough to carry two streams of traffic, narrows to single-track width after leaving the Grind, and frequent and well-marked passing places eliminate any hazards; except, of course, the impatient, usually visitors from 'overseas' un-accustomed to a leisurely way of life.

The next piece of astonishment was a magnificent hairpin bend called The Brig. The road descending turns sharply right, is carried by a bridge across water where the Burn of Roerwater tumbles into the sea, ascends steeply and disappears round a left bend, giving a sudden and fleeting glimpse of Lochend as it does so. Within minutes we left the main road, turning into a narrow way, reminiscent, except for the scenery, of an English country lane, as such things used to be. Abruptly and naturally the road ends alongside Lochend House. Beyond is the North Sea.

Stretching out before the house is a curving beach, 400 or 500 yards long, and at the far end a small hill climbs to 200 feet; on a shelf, 50 feet above the sea, is a cottage. Ours for two weeks.

2

Access is on foot. Baggage is man handled; by you, the visitor. The wise, made so by experience, pack belongings into several small cases rather than a few large ones.

Loss of a night's sleep determined that our first full day on Shetland must be leisurely and only the gentlest of exercise be taken; our immediate surroundings encouraged a stay-at-home day.

The cottage, low-built, austere looking and small, nestles in the hillside as though it had grown there, out of stone from which it is made. At the front is a central door and on either side a window, one lighting a small living-room, the other a spacious kitchen-cum-scullery. From the doorway stairs ascend steeply to two bedrooms in the roof. Furnishings and equipment are simple and adequate. There is an electricity supply for lighting and baking, calor gas for boiling rings; coal, wood and peat for burning; but no running water. This is fetched from a spring, less than 100 yards away; the Boy was delegated the duty, which he performed with much grumbling, and was ever ready to admonish lavish use.

We became fond of our cottage and named it the Little House at Lochend, subsequently discovering that its real name is Wester Haa. A footpath over short grass descends from the doorway, disappears into a gully, emerges lower down on level ground, passes behind derelict buildings and becomes part of the beach. These buildings, only the walls and gables remain, were the school, abandoned forty years and more ago. What a marvellous setting for a school, at the top of a beach washed by the North Sea.

The beach is shingle; large pebbles, small pebbles and occasional patches of coarse sand. In the spring and early summer terns breed upon it. In the late summer and autumn wading birds frequent it. Daily we saw little parties of ringed plovers paddling at the water's edge, probing amongst the pebbles to see what good things the incoming tide was bringing in; or the outgoing tide was uncovering. At night, eider ducks and their families roost upon the shingle. Twice we disturbed them as we returned to the Little House past midnight after an evening out with new-found friends. By day the beach was a never-ending source of pleasure; we too, were beachcombers, making all manner of curious finds as we crossed in the morning and returned in the evening after a day's excursioning.

At the back of this beach is a brackish shallow loch, and when the wind is unruly—it often is—the sea breaks through the shingle at high tide, cutting a channel through which it flows into the loch. Many a time we had to build a bridge with driftwood, or take off shoes and socks and paddle across; simplest was to wear rubber knee boots, and wade.

Trout lived in this loch and we had permission to fish. None were monsters; those we caught weighing from 10 to 16 ounces, just a nice size for a fish apiece for breakfast or supper.

The views from the cottage are restfully magnificent. Sharply defined against the sky on a sunny day is the ridge of a gently undulating hill; at the foot, by the water's edge, stands Lochend House, a three-storey building with dormer windows and gleaming white walls, in size and appearance more of a mansion than a house, and contrasting strongly with the crofters' little cottages scattered across the hillside. Between house and cot are cables strung on low poles bringing electricity to each; another set of poles carry telephone lines.

Between Wester Haa and Lochend House stretches the shingle beach curving round the bay on the one side. On the other the shingle gives way, reluctantly it seems, to vegetation; the loch side is irregularly edged with green, and long, grass-covered fingers poke into its waters.

On the seaward side the rounded hill top—Height of Neap—falls in a gentle curve and ends abruptly in low ragged cliff at the point Arvi Taing. Beyond, nearly 6 miles away, can be seen the southern end of the island of Yell, second biggest of the Shetland Isles, and a scatter of islets and skerries in Yell Sound.

The little headland on which Wester Haa stands, Ness of Housetter, butts into the roughly square shaped Colla Firth, sheltering our little bay on the landward side; Arvi Taing protects it on the other. Cliffs, rolling hills, the sea, a loch and a beach all invited exploration. We chose to explore the loch and the beach as these made the least claim on our energies.

Our entire fortnight could have been happily spent exploring this neighbourhood on foot; we did not neglect to do so but the car gave us almost unlimited access to Mainland, opened the way to other islands and was particularly useful as a goods carrier, with such freight as rubber boots, raincoats, cameras and fodder for the crew. Breakfast and an evening meal was taken in the cot; during the day we picnicked—on windy headlands or barren moorland, by the seashore or a lochside—and when it rained or the wind was chill, sat sheltered in the car. Food we brought at the hamlets we passed through; on a rare occasion we dined off a table in Lerwick.

Nearly always we headed south because the road to the north ends about 4 miles away at the small crofting community of Isbister. Every day brought new and unexpected wonders: still waters and wild seas; cliffs low and colourful, rugged and terrible, in fantastic shapes; sandy beaches straight and clear, another bestrewn with seaweed-covered rocks. Such a beach is at the Bay of Ollaberry.

A flourishing village is Ollaberry, reached along a minor road which ends at a cluster of houses, a garage and a store; not the big city's lordly emporium, just a compact village shop crammed with everything a crofter-fisherman, his wife and their children need; the necessities of life, foodstuffs, clothes and household goods; luxuries such as sweets and knick knacks; greeting cards, pen, pencil and paper. Ollaberry boasts a bus service too. Out to Lerwick 9 a.m. Mondays and Saturdays, 9.30 on Tuesdays and Thursdays. Back again from Lerwick at 5 p.m. 5s. 6d. single, 8s. 6d. return, when we were there.

The village church and its graveyard lie beyond the road, in a meadow, alongside the sea. The tide, ebbing fast, uncovered a smooth sandy beach studded with rocks big and small, left a mat of seaweed and small rock pools, and revealed that a flat rock, from which a couple of gulls were fishing, was not isolated but part of the cliff. For a few hours this was a golden beach revealed for our pleasure, full of strange creatures lurking beneath the rocks and in the pools; it didn't belong to man, we were intruders into the edge of the kingdom of the sea.

A mile farther on the main road—A970 to give it its official designation—is joined at a very acute angle by another road—economically this, too, is designated A970. After negotiating the hairpin there is a short, sharp climb of about 1 in 10 or steeper, and the road winds, climbs and falls, by lochs, over burns, along-

side the sea, to finish at Hillswick; another flourishing township
boasting, as Ollaberry, a post office, a bus service and a sandy bay,
much more sheltered than Ollaberry Bay because it is protected
from the open sea by Ura Firth, on which it lies, on the west
coast. The beaches offer excellent bathing and a safe playground
for children.

In Hillswick is a small factory manufacturing knitted goods
and tweeds, these products displayed in a showroom where they
can be examined at leisure and purchased. Beautiful things over
which the Girl revelled and hastily counted her strings of cash.
There were jerseys, sweaters, jumpers, lumbers, scarves, shawls,
gloves, socks and headwear. Machine and handknitted. Beautiful
woollies in Fair Isle style, hand-knitted, could be bought for 6
guineas. Later we saw them in London stores selling for 16
guineas; and gent's machine-knitted pullovers, with a hand-
knitted Fair Isle collar decoration, which the Shetlanders happily
sold for 2 guineas, were priced in London at £4 to £5. We
bought ours in Shetland, and when we wanted others, sent for
them by post.

Hillswick's scenery is ragged and beautiful. Beyond the township extends the Ness of Hillswick, its shoreline crinkled and
caved; inshore are little stacks, big stacks, inlets and skerries;
standing out on their own are the Drongs of Hillswick, resembling giant teeth but ragged, rent and torn. There are no roads;
the Ness must be explored on foot and to do so thoroughly
would take several days. We didn't. Instead we went back
half a mile and took the road to Esha Ness; it ended in a field
alongside a croft.

The coast line at this place is low, less than 50 feet, climbs
slowly, is hollowed by the sea into small caves, and reaches 200
feet. A lighthouse stands on the cliff top to give warning to
shipping of what must be the wildest stretch of sea cliff in Great
Britain, the Villians of Ure. Villains would be apt, for a more
wicked 2 miles of sea cliff would be difficult to find—gashed,
rent and torn into chasms and jagged edges by the combined
furies of the wind, sea and storm; a place that would strike fear
into the heart of the bravest sailor whose ship was driven towards
this inhospitable shore.

We made an acquaintance on a sunny afternoon of little wind, and the sea in a sober mood. The tide had ebbed leaving great expanses of rock dry, over which we scrambled and tumbled to our heart's content, until the sea came back to reclaim it. On another day we stood on the cliff head by the lighthouse, with a wind coming off the sea at gale force, great gusts of it taking the tops off the waves, throwing huge masses of water against the cliff side, where it broke and smashed with a roar, and fell like splintered glass.

Forty miles from our sheltered corner at Lochend lies Scalloway. Anciently this small fishing port was the capital of Shetland, but gave way to Lerwick more than 200 years ago when the North Sea fishing boom brought boats by the hundred, and prosperity, to Lerwick's sheltered harbour. During World War II Scalloway was the headquarters of a small band of men organizing secret operations in Norwegian waters. The full story of their amazing adventures is told in *Shetland Bus* by David Howarth.

Scalloway is on the west coast, near enough opposite Lerwick. Versatile A970 takes the motorist to both places by dividing itself into two sinuous parts, rejoining and winding its way to Sumburgh. A loop road, of mere B classification, offers a diversion for the last 4 miles to Scalloway, and another B road in conjunction with the A971, offers a longer diversion through Weisdale. In time we sampled them all. Each journey was a sight-seeing tour, never a point-to-point dash from place to place. For preference we used the A970 route; it was more spectacular. The B road lay in a valley, the A climbed over the hills, spun us round a vicious downhill hairpin bend high on Wind Hamars and revealed Scalloway on the shore below.

The view is pleasant more than beautiful, well worth lingering over. On a later visit we took a picnic lunch at this spot, and, except to avoid nipping finger ends instead of sandwiches, never took our eyes off the scene below. Scalloway is only part of it; the sea surrounding the harbour is dotted with small islands, and on a clear day the island of Foula, 25 to 30 miles westward, can be seen. We never got to Foula. The difficulty is not the single one of getting there, but the twin one of getting there, and coming back; the necessary favourable combination of wind and sea is a rare occurrence.

North of Scalloway Mainland bulges westwards, the district of Walls and Sandness, into which A970 has no access, being supplanted by A971, B and other roads. A great part is uninhabited, a waste of moorland, bog, and countless fresh water lochs, big and small. Walls, in the south-west corner, is a large village with a post office, school, a good pier and protected anchorage. Seven tortuous miles away, north-westerly, is Sandness. G. K. Chesterton's "rolling English drunkard" probably made this road for a more sinuous one I've never travelled; in this 7 miles are ninety-five, perhaps more, bends and corners. Or could it have been those "peerie folk" the trows? They lived here, we were told, and place names such as Trolligarts and Troulligarth, bear testimony to their presence.

We lingered a long while at Sandness, sitting on the rocks watching the Atlantic rollers pound the shore and the tide running swiftly through the Sound of Papa, an unfriendly stretch of water. Only a mile away is Papa Stour—big island of the priests—inhabited by a small community and worth visiting if only to see the splendid caves. We never got there.

On the eastern side of Mainland is another but lesser bulge, comprising the districts of Nesting and Lunnasting. In between the bulges are Aithsting, Delting, Sandsting and Tingwall. A 'ting', we learned, was a link with the Norsemen, an assembly of landholders met as a court to see justice done, and the names mentioned have survived from those ancient times.

We explored them one by one. Parts of Lunnasting are wild and rocky and to see them we had to take to 'other roads'; so too in Nesting, where we found stretches of coastline as wild and cruel as at Esha Ness. Aithsting, Delting and Tingwall we came to know more intimately because we were constantly passing through them to reach other places. When we ventured into these districts the day's outing was a full one and 100 miles of motoring. We passed into Delting at Mavis Grind and shortly afterwards would stop at the little village of Brae on Busta Voe for provisions if we were going east along Sullom Voe, or press on a few miles to the much more attractive Voe on Olna Firth when we went westward.

Sullom Voe is a long, winding seaway from Yell Sound and ending at Mavis Grind, where, if you have the strength of arm,

you can heave a stone into the Atlantic Ocean across the road
from the North Sea. The easterly road from Brae leads to Mossbank and Toft, from where a ferry departs for Yell. An uninteresting, desolate road, the feeling of desolation made more
acute by the ruins of a long abandoned seaplane station at
Scatsa, a relic of the Second World War; here, history records,
fell the first German bombs on Britain. We didn't return that
way but took the southerly road, a long, slow descent alongside
Dales Voe, and came eventually to Voe, round another tight
hairpin and a quick descent. The village clusters round the end
of Olna Firth, is a quiet but busy little place and has, I believe—
Lerwick excepted—a greater variety of industry than anywhere
else in Shetland. In addition to the customary fishing, knitting
and crofting, there is hand-loom weaving on a small scale and
basket weaving. Then our noses detected baking, and by following them we were led to an open window in a long building and
there our eyes confirmed what our noses already knew, that this
was a bread bakery. We saw men at tables kneading dough and
moulding lumps of it into loaf shapes. Delicious bread it was too.
We bought some in a shop, and whenever we were near Voe,
bought more. The bread made wasn't just for home consumption,
it was exported to other villages, so we probably bought Voe
bread elsewhere; but that bought in Voe tasted best.

Ten miles, more or less, of winding road separates Voe from
Aith, another prosperous little village and one having the added
importance of a lifeboat station. We like Voe better. Aith is also
established at the end of a seaway, Aith Voe, at the mouth of
which is a small island, like a glass marble in the neck of those
old-fashioned lemonade and ginger beer bottles that were so
popular fifty years ago. The island is Papa Little, according to the
ordnance survey map, but the Shetlanders call it Litla—the little
island of the priests. The map also credits the island as being
part of Sandsting, although Aith is Aithsting. Half a mile farther
on is a bigger island, Muckle Roe, and this, states the map, is
part of Delting. To map makers and Authority, Shetland is
known as Zetland, and Andrew Cluness, a Shetland man, explains
in his book, *The Shetland Isles*, that the Scots called Shetland,
Yetland, but printed their Y as Z, and in due course officialdom
called Shetland Zetland. As silly as that.

Muckle Roe is one of the smaller islands we went on to, but
not for long enough. It is an island upon which days could be
happily spent; better still, live on it for weeks; only in that way
can an island be known.

On the seaward side are wild and craggy cliffs of red granite,
and Roe, we learned, means red, hence the name; the large or
big red island. A boat offers the best view point, failing that a
long footslog over rough moorland and across the cliff tops, for
the only roads lie on the landward side serving the crofters
who live alongside Roe Sound and Busta Voe.

Two other islands we set foot on were Yell and Unst. We went
to them by the 'overland' route.

Yell and Unst, second and third largest of the Shetland isles,
are reached from Lerwick by an inter-island steamer service in
common with other northern isles. There is another route called
the 'overland', not a regular service but, when we were in Shetland, an excursion trip for holiday makers travelling by motor
coach and ferry boat. There are five stages, three by land, two by sea.

We joined the trip at Mossbank, the end of the first overland
stage which is more or less equidistant from Lerwick and Lochend. This leg of the journey passes through lovely scenery, but
as we had already travelled these roads more than once we had
nothing to lose. But yes, we did! we had to pay the full fare just
the same. At Mossbank a small boat, ferry now, fishing and
lobster boat before and after, took us across Yell sound to Ulsta
on Yell, 3 miles distant. Believe it or not but the road came with
us, or more strictly, it was awaiting us at Yell and again on Unst.
I have already mentioned the Ministry of Transport's economy
with road numbers; this road starts from Voe as A968 to disappear not to end at Mossbank, and manifest itself again, cross the
southern part of Yell, pass up the eastern coast to Gutcher Pier,
and suddenly reappear at Belmont on Unst—suddenly because
the pier and A968 (Unst) are linked by a hundred yards or so of
'other road'. What I found so baffling was that at Gletna Kirk,
2 miles up the A968 from Belmont, is a junction; then half a mile
of good road goes to Uyea Sound, a hamlet with a pier, and this
half a mile stretch has a number all to itself: B9084. Perhaps the
two sea crossings have been too much for A968; it cannot, as
rumbustious A970, carry the burden of a diversion.

No road transport awaited us at Ulsta. A lengthy wait, then a well-used motor bus arrived, turned, loaded, and wheezed off across Yell. Seen from the seat of an ageing motor bus Yell is not the most exciting of islands. We travelled north up the west-coast road which reveals more pleasant scenery than the easterly route, looked across Yell Sound to Colla Firth and Lochend, then turned inland across bleak moorland to Mid Yell. Here the island is all but cut in two, Whal Firth and Mid Yell Voe deeply indent-ing the coast, leaving a narrow neck barely three-quarters of a mile wide. Then, up and round the end of Basta Voe and across to Gutcher. A mile away is Unst, Britain's most northerly island.

Another multi-duty boat ferried us across Blue Mull Sound to Belmont; there we boarded another bus, unbelievably ancient, thirty years and more. I always thought the Isle of Man was the resting place for old motor vehicles; perhaps they sold this one to Unst.

About halfway up the island we stopped for lunch, at Balta Sound. We chose to take it in a hotel and a very indifferent meal they served, grossly overpriced; the kind of meal that makes you feel glad when it is over.

We stopped again at Haroldswick, a few miles on, so that we could buy picture postcards and post them to our friends from the most northerly post office in Great Britain; then pushed on almost to the end of Unst, finishing the run near to the top of the great hill of Saxavord. At the summit, which is 935 feet, is perched an R.A.F. establishment, a great collection of masts, radar scanners and buildings, all enclosed within a high, wire-mesh fencing—part of the elaborate equipment needed for the early detection of inter-continental ballistic missiles on their way to somewhere or other to do a lot of no good. The islanders get a side benefit. Electrical power is needed to run the station, and this is available too for the people of Unst. Mainland has its own civil power generating station. The people of the other islands either go without or harness the wind to drive electrical genera-tors. We saw many of these on Yell.

Our driver gave us two hours to see the sights at Saxavord, pointed out places of interest and left us to our own devices, not long enough to wander far afield. Two and a half miles away to the north was the lighthouse on the Muckle Flugga rock, and another half-mile away, Out Stack, which is the end of the British Isles. Below us, westwards, Burrafirth, Hermaness Hill and the great bird sanctuary of Herma Ness. Down towards the end of Burrafirth we could see the rest station for the lighthouse men from Muckle Flugga when off duty; beyond that, glinting in the afternoon sunshine, stretched the long, narrow Loch of Cliff, 3 miles of it. Westerly, the ridge of the Unst hills was sharply silhouetted against a glittering ribbon of sea into which protruded a long, dark finger of land; then more sea and faintly on the horizon more hills, probably those of Mainland at Lunnasting and Delting, maybe even further south.

All too soon we were back at Belmont, filling in the time awaiting the ferry by watching a great flock of sea-birds circling the bay, and puzzled every few moments by unaccountable disturbances in the water; something was falling into it with a great splash. Then we saw the cause: large birds diving into the sea from a height of 30 feet and more, one after another; we were seeing gannets fishing, a new experience.

Back to Yell. Back to Mainland. Back to the Little House and the little loch; and all we caught were little fish; and so supperless to our little beds. But it had been a day of great experiences.

Between the Hillswick junction and Mavis Grind the main road is crossed by one of those 'other roads'. The easterly branch ends at a spot about midway along the north side of Sullom Voe, almost opposite the now derelict wartime R.A.F.· base at Scatsa on the other side of the Voe, and has no particular merit.

The westerly branch immediately dives between two small lochs, Glussdale Water and Johnnie Mann's Loch; we never uncovered his identity—perhaps he was a victim of the trows, for this is one of their stamping grounds. The road is short and holds no promise of what is to come, but after crossing the dull, empty stretch of moorland by Trolladale Water it stops short at Gunnister Voe, a lonely spot of hushed beauty. A hundred yards back is a turning and a road that winds along this quiet voe to Nibon, ending on a beach by a few stone cots. This was a place to linger in, and we did for several hours, searching the small beaches, clambering over the low headlands and just sitting and looking. Looking at the Isle of Nibon several stonethrows away, a little world of its own with a natural arch, a stack and a skerry;

and the sea frothing at the feet of its miniature cliffs. Looking at the hill ridges piled against the sky, whole galleries of them, rolling out to Esha Ness, and rising behind them the cap of Ronas Hill in Northmavine. Then came a sharp rain shower, temporarily blotting out the distant hills. It didn't spoil the scene but added a new dimension; and when the rain had passed the hills looked fresher and cleaner, the greens were greener and the browns were browner as sun and cloud threw light and shadow across them.

The magic of it all was broken by a cry from the Boy: "I'm hungry." Perhaps he too is a trow.

Shetland is reputed to have more cars per head of population than any other county in the U.K. When motoring we rarely saw another car, but this was because we were in unfrequented places. Stopped near Lerwick we could be certain of being entertained by a continuous thin stream of traffic. We were warned that at the weekend all Shetland went motoring and were advised not to venture far afield. We didn't, but saved the weekend for visits to a couple of nearby localities, Fethaland and Ronas Hill.

On the Saturday the Girl and Boy rose early to fish. We had trout for lunch. The weather was in a give-all mood; windy, cloudy, sunny, rainy; the latter in drizzle and heavy showers, sparse in the morning, all too frequent in the afternoon when we went to Fethaland, the uninhabited northern tip of Mainland. We motored the 4 miles to Isbister at the end of the road, and then footslogged our way up and down hills, across boggy moor and rough pasture lands; 3 miles measured on the map, nearly twice that for us because of the ups and downs, the detours to miss bogs, others to see something that might prove of interest. The Boy doesn't hold with walking for walking's sake; this is crassitude. Walk to get somewhere if you must when no other means of locomotion are available; but for pleasure! on holiday! when there is a car with many a place to where that car could take you! His was an unhappy afternoon.

The highest point, Lanchestoo, 425 feet, is reached half a mile after leaving Isbister; on it is lashed a Coast Guard look-out station. The walls of the small building are stayed and the roof held down by ropes weighted at their ends, precautions taken to prevent the station from being blown sky-high by the fierce gales that rip across the heights. The ferocity of the winds that shriek across this meeting place of the North Sea and the Atlantic Ocean has to be experienced to be believed; force 8 is a mere zephyr. The fiercest gust ever recorded in the British Isles— more than 180 m.p.h.—hurled the main radar scanner off its mountings on Saxavord Hill, Unst. The scanners are now protected by specially made spherical housings, on which these terrible winds find it difficult to take a grip.

Fethaland is almost the last of Mainland; the name means fertile land. Across its green hills roam sheep without shepherds. On the high slopes are the ruins of stone cottages, refuge for sheep in storms; we too cowered behind their walls when a sudden rain squall hit us. Once these cots were the homes of stalwart fishermen who put to sea in their sixareens, big open boats driven by six oarsmen, to fish the distant grounds, sometimes not to return; more than once they were wrecked by gales and were lost in great numbers.

Gently, Fethaland slopes down to the sea, narrows to a slender waist, climbs again and reluctantly ends a half mile further on— reluctantly, because a few stacks and rocks cluster round the Point of Fethaland, slowly giving way to the sea. This is not the end of Mainland: a mile out at sea it pops up again, a long line of islets and the Ramna Stacks, a little kingdom on its own. Physically part of Northmavine, states Ordnance Survey, prosaically.

Most of Sunday we spent in idleness about Wester Haa, a sunny day spoilt by a cold west wind. Two sturdy bull calves, of about six months and known as 'the Boys', grazed on the hillside at the back of the cottage, their feed supplemented by some delicious manufactured morsels called 'nuts'. These, rattled in a bucket and accompanied by encouraging shouts, would bring them cantering across the hillside to bury their muzzles deep into the buckets—one each—amongst the nuts. They entertained us that morning and provided practice in animal studies for our cameras; the results indicated that a great deal more practice was needed. Another of their kind was a docile brown and white cow. She provided our daily milk, straight from the beast, in a bucket—none of your dairy deliveries in bottles with pretty coloured tops. She consented to a little petting and then stood quietly whilst milked. Her yield was moderate, for she was an elegant cow and not one of those proud show beasts who is not content until her udder is extended to brush the grass tops and swings clumsily to and fro like a shopping basket hung from the handlebars of a bicycle.

The sea temperature at Lochend is low; about that of melting snow, was the verdict of the Girl and the Boy. They had intended spending the afternoon swimming about the bay, but in less than ten minutes were back upon the beach hastily wrapping purple limbs and bodies in warm woollens.

The Girl suggested walking up Ronas Hill but the Boy was petrified at the thought; he had not recovered from the shock of yesterday's jaunt to Fethaland and resolutely refused to budge further than our loch. We went without him, and our sacrifice was rewarded, for he caught three trout, which made a delicious supper. We cheated on our walk by taking the car up the trackway that climbs to the R.A.F. station on the top of Collafirth Hill, another link in the early warning system. There was another in Dunrossness, that part of Mainland from Lerwick to Sumburgh. All are perched on hills, like the beacons of old which were lit to send out the message: the enemy cometh, up guards and at 'em; today the message must necessarily be swift and brief: missile coming, duck.

We left the car at the summit and walked across to the upper half of Ronas Hill, a great deal farther off than was expected. Some of the height advantage the car had given us was lost because we had to descend 200 or 300 feet to reach the slopes of Ronas. There are three hills ranged together: Collafirth, 750 feet, Roga Field, 1,201 feet, and Ronas, 1,486 feet. A long, weary climb it proved to be for our legs but well worth the effort they made as our upper parts enjoyed wonderful changing views in all directions as the slopes lifted us to new horizons. Little burns tinkled, bigger burns tumbled, sheep nibbled their way across to sweeter pastures, the sun shone in evening glory, Ronas Voe was a silver ribbon, the Atlantic shimmering gold; and the west wind reminded us that he too was out for the day.

As we neared the summit the gradient flattened, the grass thinned and our path was stony. The hill top is flat, a waste of stone and rock, and affords a wonderful panorama of the Shetland Isles: eastwards, the northern isles; westwards, wild Esha Ness and the Atlantic; southwards, immediately below was Ronas Voe, a few cottages scattered along the far shore, and then rolling hills to Mavis Grind and beyond to the lands of the Tings; northward, were the lonely cliffs and moors and beaches of Mainland's edge. We walked on and began to descend, crossed to Mid Field, which is that part of the Roga Field heights on the 1,000 feet contour, and saw, quite suddenly and 500 feet below us, a great tract of land, uninhabited moorland smothered with an intricate criss-cross pattern of lochs and lochans, chains of them linked by burns, covering 10 square miles and more, and glinting in the evening sun. We counted the lochs but were quickly confused; there are certainly 100, big and small.

The Boy, poor lad, missed it all. Just as well—those trout added the finishing touch to a perfect evening; we went to bed with contented minds and stomachs.

The road to Lerwick from Lochend is 40 miles, give or take half a mile. Never, for us, a road to hurry down within the hour, not even when it became as familiar as our garden path. We knew its moods. In sunshine, a road encouraging the traveller to loiter —like the trows of old; admire the lochs tinted by the blue of the sky; follow the example of the tiny sheep and laze by the roadside soaking up warmth, but not, as they so often did, on the road. In long shadow, a sinister road, winds blowing chill and whining shrilly across bleak moors; sheep now cuddled together for warmth and taking refuge behind rocks, bridge parapets, banks and peat stacks; and no doubt trows abroad. In storm, no longer a road but a steely ribbon of water; rain pelting stair rods and bouncing back again; burns gushing over, scattering moorland debris; hilltops deep in cloud; then, storm passing, all gold and silver, rainbows in the sky—once we saw three, one alongside the other.

No need to look at the speedometer to learn how far, come or go; we calculated time and distance by landmarks, the way divided naturally into sections. They came and went like chapters in a book.

First landmark on the outward journey was The Brig, hardly any distance down the road, with often a fisherman casting from the banks. About halfway to Mavis Grind, Eela Water, deep but rarely still, ruffled by the ever present wind that haunts the lonely moorland; alongside, the Ollaberry turning and then the Hillswick hairpin. Then the Grind itself and end of section one. On to Busta Voe and Windy Brae; across to Olna Firth and a lovely run along its banks, even better going north because of open views seaward; round a bend and there is the Voe signpost. Do we need any bread? End of section two. Five miles of utter desolation follow, uninhabited moorland piling up to 500, 600 feet and more on either side, the Long Kames—the Dreaded Kames we dubbed them, and were never sorry to see them pass and close section three.

Now the country changes, grows softer, more fertile. In the east, views of the sea, the cliffs and islets take the eye; near the road is the long loch of Girlsta and to the west, distantly, a lochan with the quaint name of Cuppa Water. Below Girlsta, eastward again, long fingers of land push out into the sea, and long fingers of water push into the land, Wadbister Voe, Lax Firth, Dales Voe. Not far from Lerwick now. A couple of miles past Lax Firth is Mainland's only major crossroad. Sometimes we stopped nearby to picnic and this was the end of section four. Only one to go; turning east at the crossroad, east again at the Bridge of Fitch, a long upward climb, across the top of the hill, and below, Lerwick, Bressay Sound and the isle of Bressay; a quick, twisty descent to the north side of the town and along the quayside to find a parking place, and the journey's done.

Lerwick is contained within a small peninsula and protected by Bressay island scarcely a mile away; between them winds Bressay Sound. Seen in plan, as on a map, the island seems to have broken off Mainland and drifted out to sea, for the two pieces are an almost exact fit. The Sound makes a deep, sheltered anchorage in which, in times past, 900 fishing vessels have moored; and the tale goes that you could walk from Lerwick to Bressay, without wetting your feet, stepping from boat to boat, so close were they.

The oldest part of the town is on the south shore where some of the houses have their foundations in the sea; built that way so that boats could come alongside and unload into the house basements; useful for smuggling, said malicious rumour.

No other capital town in Great Britain is quite like Lerwick.

Shetland. (*Above*) Our view from the Little House showing the loch, the beach and the house at Lochend. (*Below*) The great seaway of Olna Firth with the township of Voe clustered around its head

Handa. R.S.P.B. Warden Alistair Munro, leaps aboard his boat to ferry
Arthur and Ruby Willock back to the mainland

Puffins, known in Shetland as Tammy Nories and generally as 'sea
parrots'. The huge bill is coloured red, blue and yellow

The new towns have their shopping precincts which are only an elaboration in plate glass and concrete of the ancient market place, without being picturesque; Lerwick has always had one. This main shopping centre, only yards from the quayside, is paved with flags and follows the old shore line, wriggling its way north and south, between shops and houses, set at odd angles, of irregular heights. The width varies from broad to narrow, not much more than 6 feet at one point. Traffic is not barred; bicycles, motor-cycles, cars, prams, barrows and pedestrians mingle happily together at sauntering pace. There is no furious honking of horns, no impatient gestures, none of the 'out of my way, I'm a superior being, a car driver' attitude. Life winds its way up and down, over and back again, in and out, leisurely and happily along 'the Street'—officially it is Commercial Street. From it narrow lanes rise up the hillside, with houses separated by only a few feet. Above them, higher up the hill, modern houses and other buildings have been erected.

The quayside is of later date than 'the Street', built on reclaimed land. Alongside and in the harbour are fishing vessels of other nations as well as British, come in for supplies. Alongside Victoria Pier berths *St. Clair*, the pride of North of Scotland, Orkney and Shetland Shipping Co. Ltd; she sails twice a week from Aberdeen direct, and her sisters, *St. Ninian* and *St. Magnus*, call once a week via Orkney. From the small boat harbour a ferry leaves for Bressay several times a day.

On the top of Bressay, Ward of Bressay, 743 feet, we could see a high mast being erected, of familiar appearance. In past centuries the Shetlanders had fought bitterly against the tyranny of the Scottish overlords. A new menace now threatened them; this was a T.V. mast; already shops were displaying tellythings; when we came back in the following year they were enmeshed in a net of 405 lines.

South-west of Lerwick, 21 miles down the Dunrossness peninsular, is Loch of Spiggie, famed for its beauty, reached by way of Scousburgh on the westerly loop of the Lerwick–Sumburgh road; not A970 doubling, but a mere B road, excellent in all respects. We had to be very firm with our desires, which repeatedly urged us to turn aside and linger; we suppressed them, slowly and resolutely motored on, determined to reach Spiggie that day, admiring the lochs and the brochs, the hills and the sandy bays, from the car, ignoring such traps as Giants Bed and Thief's Hole set up by the roadside to delay the unwary. We did halt to take another look at Mousa, through the glasses; this square mile of island is uninhabited but has a Pictish broch, 38 feet high and the most complete of all the brochs that still exist.

The B road takes advantage of a valley at Channerwick to cross to the west coast; a most interesting road, a mile dead straight and some super winding sections abounding in double S-bends. I've often thought that together with the A970 section with which it is joined, this would make a first-class racing circuit, for motorcycles if not cars—the Grand Prix of Shetland, with a 10-mile lap.

Spiggie is reached by a metalled no-class road from the village of Scousburgh; the loch is privately owned and reckoned to be a fisherman's paradise. Our goal reached we could now linger and wander. A picnic lunch was to have been eaten on a cosy, sandy bay; having settled we then hurriedly rose, gathered our eats and found another spot, for the beauty of the place was spoiled for us by a dead and noisome seal.

Down the road a couple of miles and near the cliffs is Noss, a small crofting community, not to be confused with the island of the same name which lies east of Bressay. The crofters, men and women, were haymaking on a lovely afternoon, especially created for haymakers and the idle folk that watch. On the way down we had paused to watch two other crofters reaping a small corn field. It was done swiftly and surely with a scythe, the younger man cutting and the other, of 70 years or more, bundling, deftly tying the bundles with corn stalk. With them was a dog, a real Shetland collie, brown and black and white, a friendly little dog and, like his master, getting on in years—he was 14. The Shetland collie is much smaller than the Scottish, not taller than 15 inches, and ideally 13.

This southern end of Shetland ends dramatically with Sumburgh Head at the tip and Fitful Head on the west. From Noss the hill slopes ascend gently for a mile, easy walking, and then with surprising swiftness lift from 300 to 900 feet in half a mile, ending abruptly at the edge of a tremendous cliff wall, Fitful

Head, 928 feet at its highest, and plunging frighteningly but not sheer into the sea.

One more complete day on Shetland was left to us. We cleaned up the Little House; we began packing. We fished. We went about with our cameras. The early morning wind increased in severity. In the evening it was joined by a sudden and furious rainstorm, and put a stop to fishing, but not before the Boy had caught a last trout. Next morning the wind was at gale force. We left Lochend, battling against a wind driving hard from the south, were saddened to see familiar landmarks coming up for the last time: goodbye to Eela Water, farewell Mavis Grind. Voe was soon behind us—no bread today. Over the Lang Kames— ugh! No need to go to Lerwick, nor Spiggie. Then Sumburgh, to be told that the plane would be an hour late because of bad weather conditions but 'they' think we shall be able to take off.

There was time for a quick look at Jarlshof. We looked. Jarlshof, ". . . a unique group of dwellings of the Late Bronze Age", ". . . the most remarkable of all archaeological sites in Shetland". The name is not the original but was that used by Sir Walter Scott in his novel *The Pirate*. Here are the dwellings of ancient man of three ages, and dating from about 700 B.C. The excavations are well cared for; they lie amidst neat lawns, are there for the enjoyment of all. A tiny museum houses relics, pottery, tools, ornaments and weapons. Our visit was too brief; for the only time when we were in Shetland did we hurry. Back to the airport; a brief handing over ceremony of the car; the plane came in. At 4.30 we were airborne; at 5.30 landing at Dyce, Aberdeen's airport; by 7.15 pulling out of Aberdeen on a southbound train. But only our flesh. Our spirits were 250 miles north, in the Land of the Simmer Dim.

TWO

Birds and Other Things

---·✠·---

Everywhere in Shetland there are seabirds. Common to the coastal waters are gulls, cormorants and shags. Where the cliffs hang steep and sheer the fulmar breeds. The small island of Noss, and the great cliffs at Herma Ness in Unst, are bird sanctuaries; the ledges packed with auks, kittiwakes and gannets. On the open moorland great skuas nest in the heather. Over rocky flats and sandy beaches terns abound. Amongst the rocks and seaweeds are handsome oystercatchers.

There are rarer sights; to enjoy them you must clamber over hillsides, through heather and peat bog; then you may find whimbrel, curlew, plover and snipe. Seek the still lochans known only to the lone fisherman and sometimes a plaintive wailing will lead you to the haunt of red throated divers, who fish in the sea and nest in the quietude of the hills.

Walk softly, for these are shy creatures and jealous of their solitude.

Again we travelled overnight to Aberdeen, this time in a sleeper. The Girl was not with us, being delayed by a friend's wedding; she came days later and not liking air travel had booked on MV *St. Clair*. The Boy and I flew to Shetland from Aberdeen; and again our car was waiting, the one of the previous year, to take us to familiar places and fresh scenes.

Topping the list of what we would do was: watch the seabirds; we had chosen to go in the last week of July and the first of August, the tail end of the breeding season. Before we could begin we must return to Lochend.

Soon the well-known places were rolling past; sheep still

roamed across the road, regardless of traffic; the Lang Kames were as sullen as ever; we swung past the Voe hairpin—too late to stop for bread—and on to Mavis Grind—Grind means gateway, for us the gateway to Lochend—up and down the narrow road across the rugged moors, round the Brig—the fisherman was where we'd left him last September—down the lane to Lochend House and there was the Little House up on the hill, peat smoke curling up from its chimney; just like coming home again. Six o'clock; time for tea; the sun was shining, night was hours away. We wondered what the Girl was doing; stuffing herself with goodies at the wedding feast, we supposed.

Next morning we remembered the routine: keep off the roads on Sunday. The Boy automatically took on the role of the water bearer, uttered a grumble or two, they eased the load, and advised that water be used frugally—the spring he said with concern, looked low. He need not have worried, the rain sloshed down during our stay, Shetland's worst summer for eleven years. This day we were treated lightly—odd showers, often prolonged, morning, noon and evening.

We had never fully explored the hill above Lochend, the Height of Neap. This was as good a time as any. There was nothing of particular interest: peat hags, bogs and rough grazing, the novelty of looking at the Little House from above instead of below it, and some splendid views of Yell across the Sound. The Brig too demanded more attention than we had hitherto paid to it. Once there had been a Norwegian whaling station alongside the Voe of the Brig; that was many years ago. We had read that the site was marked by a huge gothic-like archway, made by erecting two whale jaw bones alongside the road to North Roe. The archway has gone; we did find part of it and scattered pieces of bone nearby, eaten away and pitted just as an ancient oak beam deteriorates after centuries of standing. The Boy collected a few pieces and to this day one lies on the hearth of our home, a useless, dust-collecting souvenir. Many a time I have intended to throw it away, but I never do; I know now that I never shall. Nor shall I throw away a small lump of Shetland peat that lies alongside the whale-bone. On the mantelpiece are the dried remains of a pipe fish, found on Lochend beach. We found two, and one the Boy presented to his school museum. These are a

genus of fish in the same order of the sea horse but in appearance
quite different; the body is long and thin, with bony plates; the
head is long and the jaws elongated, like the sea horse.

Exhausting The Brig we went over to the Burn of Roerwater—
it tumbles into Colla Firth under the bridge at the Brig—and then
followed the burn up the hillside, another place we had not fully
explored. Half a mile higher the burn is met by another, Twaroes,
which comes down from the west. Roerwater falls from the
north but later changes direction and comes westerly from Roer
Water, one of a chain of lochs. Above it is Clubbi Shuns then
Maadle Swankie, a group of lochans called Moschela Lochs and
a large loch, Tonga Water. All are part of the great water
labyrinth we had seen from Ronas Hill. The place was still but
not silent, isolated but not lonely. Running water tinkled over
rocks and stones; high above, a greater black-backed gull
circled endlessly, monotonously uttering its 'owk owk' cry, deep
and harsh; then suddenly and clearly came the liquid notes of
golden plover, and a brief glimpse of golden-brown plumage
as they ran from cover point to cover point.

As we crossed the beach at Lochend in the evening, a family
of eiders, a duck and her four chicks, drifted inshore, searching
the seaweed, and seeking a roosting place for the night on the
shingle; and a black guillemot was fishing out in the bay. When
we climbed over the hillock to the cot, wheatears darted before
us, flashing their white rumps. At the cot the peace of our world
was almost shattered, we really needed water; but the Boy works
to a rule: never do today that which can conveniently be put off
until tomorrow. Next morning it was raining, and he got wet;
so justice was done.

Our diary reminds us that it was a frightful day: "rain in the
night, rain all the morning, most of the afternoon and evening;
high winds and drizzle". A day can hardly be wetter than that.
We motored non-stop over familiar roads, through familiar
villages and scenery, all wettish beneath a dreary blanket of grey
sky, until our heads wagged side-to-side in unison with the
windscreen wipers. Lunnasting offered a diversion, that part of it
which pokes deep into Yell Sound was unknown to us. A place
of contrasts, empty and lonely, but not uninhabited, a wild and
rocky coast sheltering Vidlin Voe. At Lunna a 400-year-old

house, and on the north shore of West Lunna Voe, mounds, which could be the burial places of early settlers, maybe of Stone Age man. And farther on an empty, lonely stretch of water, Boatsroom Voe—but not a boat to be seen, not even an abandoned hulk. The rain eased to a drizzle, and then that too ceased; temporarily the sky had run out of rain. The scene remained gloomy: wet moorland, grey sea, and drab sky. The plaintive cry of a solitary curlew broke the heavy silence, then the rain returned. I turned the Anglia's ignition key, switched on the heater and the windscreen wipers. We comforted ourselves that tomorrow would be fine, the Girl would bring sunshine with her.

She didn't—well, not much. All too often the sun was smothered in dark storm cloud, watered down to winter strength by continuous rain squalls brought in by an uproarious and spiteful wind. Just another day spent behind swinging windscreen wipers. Three wet days in a row, and us squatting upon our sit-upons, sheltering from the rain. Then came a golden day, and, as the Girl's inside was still tender from her night with the wild North Sea, we remained at Lochend, exploring again the never-never land behind the hillhead and beyond the bend of Arvi Taing. Yell Sound was sapphire and silver, the hills emerald; rabbits basked at the doorways of their burrows, the wheatears' rumps were whiter than white. A wee snipe escaped from its imprisoning egg, dried itself in the warm sun and set out to explore a green paradise, knowing nothing of greater black-backed gulls and hoodie crows, or of wind and rain; he (or she) copped it on the morrow. The trout were more wily, refusing the invitation to grace our table; only a small flounder got hooked, and that was returned to the water as much in pity as in disgust.

Fishing our loch brought nothing but tiddlers and disappointments. Nor were our ventures elsewhere any more successful. Eela Water, deep and wide, contained nothing more foolish than a solitary 3-ounce trout. The Boy was convinced that the Burn of Roerwater was a great storehouse of fish. Its waters came tumbling and frothing down the rocky moorland in long stairways, pausing at intervals to form deep pools, as if to recover their breath. In these, the Boy insisted, were trout, and in his mind's eye pictured gargantuan specimens falling to his wiles. He was proved right, there were trout; he and the Girl fished a number of these rocky cisterns for several hours and were constantly rewarded with bites—nips, more factually, none of their catches were more than 4 ounces.

I left them to it. I am not an ardent fisherman, have not the patience; I am much happier with a camera filming elusive fowl. On this afternoon I was seeking the golden plover. Not a glimpse did I get, not a trill did I hear. I left Roerwater at its confluence with Twaroes and followed the latter as far as Queina Waters, a steep mile upstream, and then struck northwards across flat terrain, meaning to reach Roer Water.

North-east and north-west the moorland is studded with innumerable and nameless lochans and burns, none that I saw on this dull and sunless afternoon of any particular beauty or merit. Just water in an almost empty world. With unexpected suddenness a strangely weird and plaintive wailing interrupted the tired sighing of the wind and the faint croaks of a high-flying greater black-backed gull; not once, but at quick and regular intervals, without ceasing. I listened, motionless, until I had gauged the direction of the sound, set off slowly and softly, and within a few minutes came to a narrow lochan, sunk deep in a rocky basin. On the still waters floated two birds, side by side. They were about the size of mallards, their heads were grey and the upper parts grey-brown, and on their necks was a dull red patch. A pair of red throated divers—not uncommon in Shetland, but you must walk the hills if they are to be found in their haunts. We had seen them fly over Lochend, and saw too their big brothers, the great northern divers, who do not stay to breed.

This pair of red throats were unconcerned about my presence. They swam slowly down the lochan to the far end, turned and came back, and turned again. One, and always the same one, would lower its head a few inches every minute or so, thrust out its neck and utter a mournful wail. Occasionally, together or separately, they would submerge their heads; from time to time, suddenly and without apparent effort, a bird would sink beneath the surface, disappearing for several seconds—up to twenty I guessed by simple counting—and reappear yards further on. Their unconcern encouraged me to erect tripod and camera. Co-operatively they continued to swim and dive. Hoping to approach near for big close-up shots, I casually moved round to their side, and with equal casualness they moved across the water so that we merely changed places. We played this game for an hour, and then I gave up, intending to leave them to their solitude. Quite unexpectedly they left me. Their departure must have been caused by a movement I made with the tripod; the legs are polished aluminium, possibly caught the light, and frightened the birds—I was watching them instead of my own movements. With incredible swiftness one bird rose, was off the water and away into the distance, flying powerfully in a matter of seconds, whilst the other sank instantaneously. I grabbed the camera too late to catch the bird in flight, and waited anxiously for the other to surface. When it did the camera wasn't covering that spot. The bird seemed to come straight out of the lochan into the air and was gone, wings beating powerfully. I caught it as it lifted off the water and flew rapidly after its mate.

We found another pair of red-throated divers days later when on the way to Scalloway. A couple of miles past Voe the road skirts Loch Gonfirth and as we swung round the bend I spotted the two divers swimming. We halted, walked over to the loch to watch, but the divers were agitated, swam away, rose quickly and flew across the hills. The car passing by did not disturb them; three humans standing by the loch side was too much. Hunting lonesomely has its disadvantages. Occasionally a rarity is found, but unless there is the evidence of another pair of eyes in corroboration you must cherish your find alone, the world cannot share it with you.

No more than 3 miles beyond this great water labyrinth of North Roe is a coast line of wild and rugged beauty: little bays and headlands, stacks and skerries, natural arches and caves, hills that slope steeply to the cliffs; no habitations, no roads. We reached it from the village of North Roe, taking the narrow by-road to Sandvoe where it ends, as so many roads do in Shetland, at a beach—a quiet sandy beach, much more wholesome and pleasant when the tide came in.

For 3 miles the coast faces north, then turns at Uyea to face west. The cliffs are rugged but not spectacular, rarely climbing higher than 50 feet and eroded into tiny rocky bays. Eiders were escorting their chicks in and out of these bays, keeping close into the cliffs, never taking a short cut across. Two families accompanied us for a long time, feeding as they went; we stopped and made a meal of it. Further out at sea cormorants were fishing, swimming low in the water, necks erect, bills uplifted. They dived frequently, sometimes disappearing by sinking like a submarine, but more frequently with a quick jerk. After a fishing spell they would fly ashore, alight on a rocky perch and, ever watchful, preen and dry their feathers in the wind, standing, occasionally, with wings outspread, eagle fashion. Two flew up to a headland no more than 20 yards below where I was sitting alone, and stood there for many minutes, continuously alert, heads twisting and darting restlessly, but ignoring me. I edged closer slowly, feet at a time, camera in hand, taking shots at every move to ensure that I had something on film. Eventually I was close enough to get them full frame; fear of toppling into the sea dissuaded me from attempting a nearer approach, for the hillside is steeply banked. I lay on my tummy mostly, using small and convenient rocks standing proud of the grass, some only just, to steady the camera; erecting a tripod might have scared the birds.

At Uyea there is a small, low plateau, and round this the coast makes a 90-degree turn; opposite the corner is an islet, also called Uyea and very close to Mainland at one point. The narrow channel is strewn with rocks. It may be possible to scramble over them to Uyea, but except for the achievement and the view of the Mainland then obtained I cannot think of a reason for doing so—for some that is reason enough.

Further down the coast, now facing west, we came to high cliffs and wide geos. At the bottom were rock slabs and wide ledges, some canopied, the lower slabs washed by the surging seas. On the dry slabs cormorants were nesting; we could see chicks in the great seaweed nests, and there was a continual coming and going of adults. The upper cliffs hang steep and sheer, earthy in places, clothed in green, and on these upper ledges fulmars were nesting. Very few of the breeding sites were occupied by adults, just a chick in down, a great fluffy ball twice as large as the parents. The adults were flying, as fulmars will, stiff-winged up and down the cliff face, round the bay, out to sea, ceaselessly, tirelessly, just for the joy of flying.

From where we stood above there appeared to be a way down

to one of the ledges occupied by a nestling: a circuitous and pre-carious way with a long steep drop, smack on to the cormorants below in the event of a false step.

There was no direct route down the cliff side, I had to make a zig-zag path. The first zig was easy, no more difficult than lopping down the garden to pull rhubarb; but the first zag was strewn with difficulties. In places the ledges were earthy and soft, merely inches wide, offering only a precarious foothold. Controlled downward progress was possible because just above head height were a few conveniently placed rocks on which a secure finger-hold could be taken, although at the expense of torn finger nails. The second zig was deceiving and twice I had to turn about and choose alternative ways. That which looked simple and straight forward from the cliff top was alarmingly impossible when reached. Patience, common sense and method were what was needed, so I talked myself down. Gingerly take a few steps, crouch and duck under that protruding rock, sit, slither down the next few feet until the step is reached, shuffle forward on sit-upon, feet down, toes outstretched, when they touch ground, gently ease forward and stand upright; if they do not touch, try and gauge distance, slither forward and downward until they do; come upright and regain balance, pause for breath and remember that this is really enjoyable; if balance not regained—mad to have come down at all, bad example for the Boy. There he was at the top impatiently waiting, 15 minutes had passed since I had left him. He grumbled when he was left. Envy, I had to remind him, is a sin; and, to clinch the argument, asked him, "Supposing you were to fall?" "Supposing you fall!" was the rebellious answer. I handed him the second ciné camera and instructed that if I did slip he was to start filming and keep on filming, not to come to the rescue until I was lying still. Judging by the way he kept the camera trained on me he was evidently very mindful of that instruction.

The second zag was short and easy; the last zig unexpectedly difficult. A raincoat, a camera in its case and slung over one shoulder, and a tripod, hampered free and easy movement. The tripod had to be lodged on any nearby tussocky growth, ledge, or rock ahead of me, retrieved as I passed, and again passed on ahead. Then, when only 20 yards from the nestling fulmar and with an apparently straight but narrow path ahead, I was trapped. Completely. I came to what was, to all intents and purposes, a steep rocky path, 15 inches wide and offering a firm, safe foothold; when I reached the end I discovered that I stood nearly 4 feet above a ledge on to which I must drop. Too dangerous, I decided, and turned to retrace my steps, only to find that this was a one-way approach, there was not a way back. That comfortable 15-inch ledge shrunk horribly; below it was a steep, earthy slide; at the bottom rocks. I looked up, straight into the lens of the second ciné camera; on the button was a grubby finger.

There was only one way out of this trap, to go forward backwards; I sat down on the ledge and tried turning over on to my tummy; my rain coat would not twist with me, the camera was in the way, the tripod an encumbrance; so these I dumped on the ledge behind me; then turned slowly on to one side until my sit-upon came up against the cliff face and tried to push me off; by turning over the other way round my sit-upon simply hung over the edge of the ledge, and there I was lying on my tummy. Now to wriggle down the ledge and get my feet over the end. A lot more wriggling and my knees were over. The next move was tricky. I had to wriggle gently until my waist reached the edge, then I could bend my body so that legs hung downwards. Then came a moment of minor panic, I could not feel anything solid beneath my feet. The lower edge could not be more than inches away so I wriggled a bit more, and relief—my feet touched solid ground. I stood up, grabbed coat, camera and tripod. The rest was straightforward.

The fulmar chick was unsociable and opened its beak. Chicks, I had been told by other people's experience, can spit oil too. This one confirmed the fact, firing volley after volley, and when its oil was exhausted, vomited its breakfast, and stained its snowy white downy front browny yellow; then sat back and hiccoughed, or the bird equivalent.

The beginning of the menacing display was simply a backward movement of the head and a slight opening of the bill, not to full gape. When the first stream of oil was ejected the chick was on its feet, leaning backwards on to its tail stump, probably to support itself as the young legs were wobbly. Then—whoosh, and the tripod legs were smothered, and that from a distance of 4 feet.

The next shot fell very short, and then there were just dribbles as the chick settled down, more or less, and consented to be filmed.

The way back was simple and straightforward. Within ten minutes I was back with the Boy, who complained that I had been a long time. I had, too—more than an hour—all for a few film shots of seconds' duration.

Fulmars have established colonies on most of the Shetland Isles, with the biggest concentrations at Herma Ness and on Noss. Those we found on Mainland were small, up to fifty pairs, and the one in this small, north-western corner had only twenty-five pairs. Occasionally we found a solitary nester; we had one of our own at Lochend, near the headland of Arvi Taing. Our noses discovered it before our eyes; the musty odour is unmistakable. Although the nest site was low, less than 50 feet above the sea, it was in a typical setting, a small bower partially hidden in lush vegetation. The fulmar shuns the bare rock cliff so beloved by the razorbill and guillemot, and in Shetland has even forsaken the cliffs, nesting inside tumbledown cottages, a surprising tendency because the bird is making difficulties for itself. In the air the fulmar is a graceful creature, but on its feet the most clumsy and takes to the air with difficulty. From a cliff it can launch itself with not much effort. Come to a nest unexpectedly and the disturbed bird scrambles off as though it has no legs to walk upon, shuffling forward not on its toes but on the tarsi. I once watched a ringer operating on a fulmar colony on the island of Dùn of St. Kilda. The birds were nesting amongst tall and thick vegetation and when disturbed shambled forwards, using their wings but unable to gain height. Most of them came down, entangled in the vegetation and were easily caught; they spat furiously—how that man stank!—were quickly ringed and tossed into the air, and returned to grace.

In the air the fulmar is superb, whether it is skimming the waves or soaring up a cliff; in comparison other birds are seemingly without grace, eclipsed. We watched their almost effortless displays in a gale by the cliffs at Esha Ness, near the lighthouse, the wind coming off the sea in tremendous gusts. Pipits searching for food in the grass and pools were blown off their feet. The razorbills and puffins were just so many black and white shuttlecocks. Courageous fliers these little birds, for time and time again they fought back against the wind until they reached their ledge and burrow. The gulls looked uncomfortable and the oyster-catchers preferred to probe the rocks. The fulmars were masterly. By this time of the year, early days in August, most egg hatching is done, the young have left the cliffs and are at sea. Excepting fulmars we saw few chicks. The nests of the land birds are not easily found, and to search aimlessly is time wasted. Often we saw pipits and wheatears carrying food in their bills, and watched them closely, but never found the well-hidden nests. The wheatear is most elusive; I have never come across a nest—nor that of a lapwing.

That long-legged wader, the oystercatcher, is very common in Shetland and was present in large numbers on the moors of Northmavine, our constant companions, noisy and excitable. The chicks—they leave the nest within hours of hatching—are mottled brown and not easily seen against a background of rocks, shingle or moorland, so good is their protective colouring. Added to that, the chicks freeze, remain as still as their surroundings, as soon as the adults cry alarm.

In Shetland oystercatchers are called shalders. Shetlanders have their own names for many other species of fowl, some much more attractive and picturesque than those by which they are commonly known: thus the great skua is called the bonxie, a young gull a scorie, the greater black-backed gull a baagie, the black guillemot a tystie; a fulmar is a maley, a puffin a tammi norie, an arctic tern a tirrick.

Besides our fulmar at Lochend, the wheatears and a few herring and greater black-backed gulls, we had eiders, occasional cor-morants who came to fish in the bay and preen afterwards on the skerries, a lone black guillemot which we suspected might be nesting somewhere in Colla Firth, a flock of starlings and a few blackbirds. We found disused blackbirds' nests in the walls of an old building, that had once been a byre; and the starlings, too, nested in old buildings and walls just as they do on Hebridean islands. There was a family of starlings in the chimney stonework of the abandoned school buildings below the Little House. The adults were very apprehensive and refused to enter the chimney if we were watching or stood near the building, flying around from perch to perch, luscious grubs hanging from their beaks.

This behaviour first attracted our attention; then, hidden from the parents' view, we watched them go to the nest. The parents used different routes, one always entering from the east side and the other from the west. As we walked past we would hear the young birds chattering. On one occasion there was the ludicrous situation of the parent birds refusing to go to the nest for fear of revealing its position, so flying around carrying food in their bills and trying to appear as though they always did this just for fun, and a young starling hanging head and shoulders out of a crevice in the stonework, voraciously demanding food.

The eiders would swim twice around the bay. In the morning we would see them going east to west—two ducks and several chicks; in the evening they would return, swimming west to east, at about dusk, foraging amongst the seaweed, and quite often would camp for the night on the shingle of our beach.

The gulls, no more than twenty, were mostly to be found about half a mile away on or about the small loch of Housetter, idling their time away swimming, feeding, flying and bickering. A motley crowd of adult herring gulls and a few second and third-year birds, and three or four greater black-backed gulls in adult plumage.

Our diary of the weather makes dismal reading; day after day the rain clouds stormed across the sky causing us to spend more time motoring than walking. Tiring of our car, we abandoned it one day at Scalloway and spent an afternoon afloat cruising amongst the western isles, all in calm waters, the master of our vessel assured us, and saw Shetland from a different angle. For three glorious hours we weaved in and out of the skerries and islands, passing islets such as the Sandas, big rocks really, topped with grass and heather; bigger islets like Oxna, Papa and Hildasay; and the big islands of Burra and Trondra.

Burra is almost two islands lying one behind the other and joined, like siamese twins, by a short isthmus on which a school has been built. West Burra is the larger and has a flourishing village at its northern end, Hamnavoe, and a well-protected harbour in the Hamna Voe, sheltered, too, by the islets of Oxna and Papa not much more than a mile away. The village clusters round this harbour, and the villagers are mostly fishermen. We called on our way back to Scalloway to pick up a few passengers. The other part, East Burra, is shorter, a narrow sound separating it from

Trondra, to the north. Trondra is an island of low green hills, and sparsely populated. Its other end curls round Mainland, protecting Scalloway harbour.

Many of the skerries and islets are in the possession of seabirds. On the rocks and crags stood shags and cormorants, clouds of terns hung over beaches, fulmars skimmed low over the placid water, and an occasional black guillemot would dive under our bow. On flat rocks, basking in unexpected sunshine, were seals. We sailed close inshore to Mainland to see the cliffs at Rea Wick; these headlands, out of view on land, offer a wonderful spectacle from the water and, as always, are monopolized by the seabirds.

A wonderful three hours. Apart from one shower the sun shone with summer warmth and the waters were as calm as promised. When we went ashore Scalloway was preparing for an evening regatta. We stopped to watch racing begin, then headed northwards, and no sooner had we climbed to Wind Hamars than the sky turned black and a storm of unexpected ferocity lashed us. Windscreen wipers were useless in such torrential rain, so we halted until it abated. On the hills between Aith and Voe there was a second storm, which ended as abruptly as it began, drifting eastwards. Along the banks of Olna Firth we halted to look back; hanging in a black sky were two rainbows. Then the sky lightened, the sun shone brightly, one rainbow began to fade and as it did so a third began to glow.

Three times we attempted to set foot on Noss, twice we failed; the weather must be favourable before a landing can be made. The trip is a four-stage one: to Lerwick, ferry boat to Bressay, cross the width of Bressay—two miles, over the Sound to Noss. On the mood of this 250-yard stretch of water depends success. Noss Sound is short, 600 yards at the most with a funnel at the northern end, and when the winds are in a certain quarter the water comes hurrying through the Sound and makes the crossing unpleasant and hazardous.

The Lerwick–Bressay ferry runs frequently, meaning several times a day; the trip is a short one, a little over a mile and worth making because of the wonderful views obtained of Lerwick. On the ferry crossing Bressay is not shown to advantage. The west coast is unimpressive, the southern end and the east coast are the most appealing. Our first attempt on Noss was thwarted by un-

Handa. After a long day in the field The Boy prepares a late evening
meal in the bothy

Handa. The Great Stack, rising 300 feet sheer from the sea and
colonized by thousands of nesting birds

BIRDS AND OTHER THINGS

favourable winds and seas so we walked southward along the
west coast of Bressay to the lighthouse, a dull 2-mile walk which
roused the Boy to recite an unending list of the better things we
could have done. The lighthouse and cliffs at Kirkabister Ness
revived his flagged spirits temporarily, but he resumed his
soliloquy on the foolishness of walking, on the return. For once
we agreed with him, it was dull.

The second attempt was foiled by a gull. To avoid a further
disappointment after a 40-mile trip to Lerwick I intended to
phone through and ask if the crossing was possible on this
particular day. Lochend House's phone was out of order, quite
dead. So, too, was the public call office at Lochend post office a
mile away. On the return to the Little House I saw that the
phone wires, overhead open line, were twisted together in a tight
bunch; hence the discontinuance of phone service.

The cause was an immature herring gull, one of our local birds,
which had a wing entangled in the wires. Frantic struggles to free
itself had only worsened its plight, for it had fluttered round in a
circle catching up the other wires until bird and wires were
tightly bound together. The gull was now dead, hanging from
the wires by its smashed wing. Efforts to dislodge it with a pole
were unsuccessful, so enmeshed was the bird's wing and the
phone wires. The next nearest public call office was several miles
away and from there I was able to tell Lerwick of the breakdown
and at the same time learnt that Noss Sound was this day navig-
able. Too late! midday was past, we would have landed just in
time to re-embark.

At the third attempt we got ashore in company with a score of
other folk. The Bressay ferry was met by what was euphemisti-
cally called a taxi. Philologically correct, but not the brightly
gleaming motor cab, licensed to seat four, that whisks travellers
to railway stations; just a motor vehicle into which the driver
crammed as many passengers as could get in. He crouched by his
wheel, not in splendid isolation but sharing his worn bench with
two passengers, and these were the most comfortable seats of all.
In the rear, a windowless body and open back, were benches;
when no one else could possibly get in, however hard they
pushed, the taxi was deemed to be full and rattled off across
Bressay. The road ended as a rough trackway at a cliff head. Here

4

the passengers disgorged and descended a fairly gentle and rock-strewn path to the sea's edge. The taxi picked up returning visitors and went back across Bressay.

Amongst those with whom we travelled were two women school teachers of indeterminate age, agog with excitement at the prospect before them. There was a noticeable falling off in their enthusiasm when Noss came in sight. The path from the hillhead gradually merges into the surroundings the nearer it approaches to the sea, and finally is not there any more—at no particular point can it be said that this is where the path has ended. The last 100 yards or so are rock and seaweed, over which we slipped and scrambled until a large rock was reached, against which the sea lapped. This was the place of embarkation. To this rock the shepherd from Noss brought his boat, into which his passengers stepped, slipped or fell, according to his or her dexterity and nimbleness, then he rowed them across the Sound and disembarked them upon another rocky sea-washed shore. Our two school ma'ams never reached distant Noss. One, less nimble than the other, slipped on seaweed on Bressay's shore, and sat down slap in a rock pool, one of the deeper kind. The more fortunate of the pair, full of dismay, helped her companion back on to her feet, uttering sympathetic coos; there was little else she could do; there was not any dust she could brush off. Fallen madam was bruised and damp, and although her skirt was a thick one I reckon that her panties beneath were uncomfortably wet. The one made excusing noises for both, abandoned the expedition, bade the boatman not to wait, and went away, I supposed, to dry out. There was a good wind blowing.

From the landing place Noss climbs from sea level, gently at first and then abruptly to 592 feet in about a mile. The covering is heather and grasses amongst which the bonxies and the arctic skuas nest; so the walk across is enlivened by their dive bombing. The attacks are intimidating, not formidable nor dangerous; occasionally a hat may be knocked off and there is, I suppose, a million to one chance that scratches could be inflicted or, say, the loss of an eye. This is a risk which must be taken when trespassing upon a breeding ground. The timid should wear glasses, rosy tinted.

Noss has two show pieces: the Noup at 592 feet which falls sheer to the sea; and the Holm, a great pillar of rock standing 70 feet distant. Both are colonized by seabirds, an amazing spectacle assaulting three senses; sight, hearing and smell. Where there's a ledge there's a seabird. If it isn't occupied by a razorbill then a guillemot, puffin, kittiwake, gannet, fulmar, gull or shag will have it. There they stand in serried ranks, nodding, shuffling, squirming, bickering. The sight is unbelievable, the noise incredible, the smell overpowering. A trip round the island by boat is the best way of appreciating the cliffs and the birds. This was a trip we never made. Our stay on Noss was limited to brief hours; a fortnight would not be too long. One day I must go to Noss again, and take my tent.

The weather never relented enough to give us another day of complete sunshine. Our diary of our last-but-one day on Shetland reads: "remained at Lochend, wet morning, some sun in the afternoon, then rain and mist". A sea mist which hung a grey veil over everything including our spirits. In the evening we revived them with a farewell party, and stumbled across the beach, back to the Little House, at 1.45 a.m. to the dismay and disgust of the grumbling eiders.

Next morning was misty too. At eleven rain began, and so it continued throughout the day. We were all travelling back together, by boat, on MV *St. Clair*, as I thought I would like to sample the sea trip, and air travel makes the Girl sick—so does the sea. We left Lochend in drenching rain at 1.40 p.m. and, as there was nothing to linger over, were at Lerwick in 65 minutes. Our luggage went aboard, we were to sail at five. We handed back our car, squelched around Lerwick, and went aboard to stay.

Five o'clock, sailing time. We took a final look at Lerwick. The Lerwegians crowded the quay, waving goodbye to friends and relations, and the gulls lined the roof tops. Slowly, *St. Clair* left her berth. In twelve hours more we would dock at Aberdeen. We left the harbour escorted by the gulls, and slowly, slowly slipped into Bressay Sound.

By plane the departure is swift, there's no time for retrospection; by boat, leave-taking is a leisurely affair. Slowly Bressay lighthouse receded and Mainland's coast was almost lost in the mist. Gradually it grew fainter and was gone; we were in the open sea. A lovely holiday should have no definite end, just fade away, like a beautiful dream.

THREE

The Beginners' Isle

———————————— ⋯✤⋯ ————————————

Its name is Handa, a corruption of Sandy, an island of sandy beaches and rabbity dunes; a wild place of heather, bracken and coarse grazing—and Torridonian rock. It lies inshore, half a mile from the Sutherland coast in north-west Scotland, is often rain-swept, whipped by the winds and battered by the seas. On the landward side a moor sweeps down to a reefed shore; seaward, rusty sandstone cliffs rise sheer from the Atlantic to more than 400 feet.

Man deserted Handa more than a 100 years ago. A few tumble-down shells of cottages remain. Time and unrelenting wind and rain have almost obliterated his mark.

Handa could well be called the feathered isle, for it is the breeding haunt of some 100,000 sea fowl who come, year after year, to colonize the sea cliffs: 60,000 guillemots, 4,000 fulmars, 500 shags, 12,000 razorbills, 800 puffins and many others in numbers great and small, to nest and breed upon the rocky terraces, tier upon tier.

Late in the afternoon of a Friday in mid-June, and a perfect summer's day, we—the Boy and I—pointed the bonnet of our car to the north and let in the clutch. Seven hundred and fifty miles of road and a mile and a half of Atlantic Ocean separated us from Handa Island.

Three hours later we joined the Great North Road near Huntingdon. The Boy said, laconically, "There's no oil pressure." Nor was there! We stopped, but there was nothing we could do; so we restarted, and the engine emitted expensive noises.

Three more hours and we struggled into Newark, under cover

of a smoke screen. The time was after ten, and dusk had fallen. The engine needed expensive repairing, and we were past caring; so we went to bed, to sleep on the problem; how to get to Handa, now 600 road miles and one and a half sea miles away.

The answer was simple, of course. We sold our car for what it would fetch, hired another car, and continued. But my heart broke to see that car go. Such a wonderful motor-car—a Mark 7 Jaguar—a thirsty brute at 18 miles to the gallon, but it would top the ton with ease, cruise at 90 and hold the road like a leech. I had just fitted four new tyres, a battery and a brake master cylinder. Engine repairs would have cost more than the market value of the car; and we were stranded.

In Newark we found one of those kindly concerns that will take a broken motor-car off your hands.

"How much do you want for it?" asked the man.

'I'll take £25," I began.

He shook his head sadly.

"How much?" I asked, hopefully.

"Fifteen."

"That all?" incredulously. "The tyres are new. . . ."

"That's all," he confirmed.

I accepted reluctantly and unpacked the Jag. There were fifteen packages: food, clothes, photographic equipment and sundries in elderly suitcases and cardboard cartons. Surprisingly, all went into the hired car although it was only a half-pint model —an Austin A40.

We were now well behind schedule, for we didn't leave Newark until Saturday noontime, the hour at which we should have been crossing the Border.

The A40 proved to be a very gallant motor-car which could be driven hard hour after hour. The speedometer needle was hovering at eighty more often than not, a probable true speed 70 to 75 m.p.h. Remarkable was the petrol consumption; despite continuously hard driving, the average was 40 m.p.g.

After we left Gretna, late in the afternoon, rain set in and stayed with us until we reached Callander at eight that evening. There we spent the night; left soon after nine next morning, with 275 miles to go, and through country that was entirely new. We made good time and were able to laze awhile in the hills above

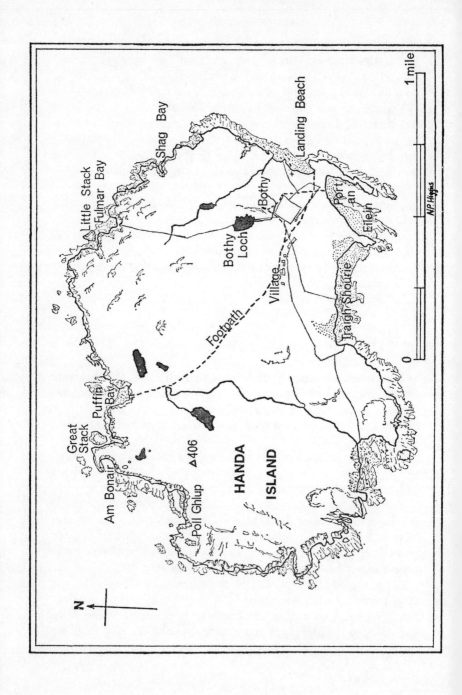

N

Shag Bay

Little Stack
Fulmar Bay

Landing Beach

Bothy

Bothy Loch

Port
an
Eilein

Village

Traigh Shourie

Footpath

Great Stack

Puffin Bay

Am Bonair

Poll Ghlup

△406

HANDA
ISLAND

N.P. Higgins

1 mile

0

Loch Leven before moving on to Fort William. Thence by that wonderful Grand Prix road alongside Loch Ness to Fort Augustus, across to Beauly and over the hills to Lairg. With only 50 miles to go we made a lengthy stop above Lairg before motoring up Loch Shin. At eight o'clock that evening we rolled down the hill to Port of Tarbet and parked the Austin on the tarmac by the jetty. The journey had been a long and pleasant run, but we were at the point of hating motor-cars. Across the Sound we could see Handa Island, and smoke rising from the bothy chimney.

That night we spent with Alistair Munro, the R.S.P.B. warden of Handa, and next morning he put us on the Island along with our fifteeen packages—food and film for ten days.

A map of the island, scale 6 inches to the mile, which the Scottish office of the R.S.P.B. supplied for a modest tanner, shows that the bothy is but 400–500 yards from the landing beach. What it does not show is that the first 300 yards are sand dunes, which rise almost vertically, thus ensuring that however cold the crossing, which takes about fifteen minutes, he and she arrive at the bothy in a muck sweat.

Half an hour after we had landed, rain set in. There was no abating; it fell monotonously, drenchingly for seven hours and then intermittently for several more. Nothing to do but get settled into the bothy and acquainted with our fellow islanders: Ruby and Arthur Willock from a southern part of England that lies between Oxford and Reading, and Mary and Cy Hampson from Alberta, Canada. The Willocks were going ashore three days later and the Hampsons the day after.

By mid-afternoon the Boy and I were restless. We were on holiday and came to see Handa, explore and enjoy it; and all the rain in the world wasn't going to stop us. At that moment all the rain in the world seemed to be concentrated on Handa, so great was its ferocity. Before we left, ten days later, we were to know worse. June, weathermen say, is a dry month in Scotland; subsequent years showed that there are grounds for this, then surprising, statement.

The Boy elected to go fishing whilst I set off to squelch across 500 yards of bleak soggy, moorland and flounder through peat bog until I reached the firmness of a rocky shore at Am Bonnaid, which is just below Shag Bay. There the cliffs are insignificant but rapidly climb and within a mile have reached nearly 400 feet at Puffin Bay and the Great Stack.

Along this short but rocky coast, indented by small, boulder-strewn bays enclosed by rocky heights, is a tremendous concentration of seafowl, lending their names to their most favoured haunts. Thus we have Shag Bay, Fulmar Bay and Puffin Bay. But the hub is the Great Stack of Handa colonized by thousands of nesting birds and rising 300 feet, sheer from the sea. At its nearest point the Great Stack is separated from the island by only several yards, but is quite inaccessible to these who are earth-bound. Not so the seabirds; it is their haven and their citadel where they dwell tier upon tier: kittiwakes just above the sea, then guillemot slums and razorbill tenements, the grassy summit fringed by puffins and patrolled by their natural enemy, the greater black-backed gull.

Five hours it took me to cover that royal mile of birds, there were so many to watch despite the lack of activity—even a sea-bird must surely tire of rain. There was so much to see, so many little bays to slither down to, so many headlands to scramble over; a full-scale cliff colony of thousands of breeding seabirds is enjoyable whatever the weather.

About seven the rain eased to a drizzle and an hour later I reached the Great Stack—roughly 350 yards an hour. Ten days later the journey took us as long, there was always so much to watch and see and check. On this first occasion, mostly there was general and commonplace activity. All the colonists lined up for routine inspection: guillemots wriggling, razorbills nodding and shuffling, puffins notably scarce, probably—and sensibly—deep in their burrows, sheltering from the rain. Several shags stood guard by their nestlings; one adult flew off the nest to reveal a lonely egg. Fulmars squatted on the ledges, which might have been nests; some sat regardlessly by their egg; and one was either building a nest with fresh, green vegetation, or decorating a finished one. An artistic bird this, for there is no more spartan nest than a fulmar's: a rocky ledge, or shallow, bare scrape on an earthy cliff, a few stones to mark the boundary, and eventually an egg to indicate where to sit.

The cross-country journey back to the bothy, enlivened by the sight of four lapwings mobbing herring gulls, and two involuntary and knee-deep excursions into boggy patches, took but an hour. The Boy was awaiting me; there weren't, he said, any fish in the Sound.

Judge our disappointment when next morning dawned wet! Slowly—oh so slowly!—the skies cleared and the sun was left in charge, spreading mellow warmth, hinting that summer had come to Handa. It was not the best of days; the clouds again piled and the evening brought intermittent rain.

We were abroad nine hours, exploring sandy beaches, reefs and low, rocky cliffs along the southern coast. Across the Sound, a mile and a half away, is the mainland, the barren hills of Sutherland rising green and grey and brown, and hazy mountains to the north and south stand sentinel, aloof.

Along this southern shore of the island life is more peaceful than in the teeming metropolis of the north. On the reefs by the shore eiders secrete their nests, ringed plovers scrape sandy hollows on the beaches, meadow pipits weave their dainty cups amongst the tall grasses, razorbills shuffle and nod to each on low cliffs, and herring gulls scream incessantly at human intruders on their territory.

Above the little sandy bay of Port an Eilein is the old burial ground. No stone walls mark its boundaries as on Hirta of St. Kilda and Tanera Mhor of the Summer Isles; just a few grass mounds, rough and weathered headstones—except for one, without inscription. The exception is a horizontal slab lettered "Peter Morrison". An islander, perhaps, as it was the custom of the mainland folk to bury their dead on islands so that their remains might lie peacefully, undisturbed by foraging wolves. Thomas Gray might have written his lines especially for them:

> Where heaves the turf in many a mouldering heap,
> Each in his narrow cell for ever laid,
> The rude forefathers of the hamlet sleep.

Not so very far away the ruined cottages of the Islanders stand alongside the pathway from the bothy to the Great Stack; heaps of tumbledown stone, overgrown with grass, nettle and bracken; some with walls standing shoulder high; here was a doorway, and there a fireplace where peat glowed warmly.

On this quiet sector of the island we spent a happy afternoon and evening. Inshore were a dozen eider ducklings bobbing and diving under the watchful eyes of three ducks, and 100 yards out in the Sound a pair of shelducks swam purposefully across the bay. When we returned that way in the evening the eider were still at play and the shelducks were accompanied by three chicks.

We lazed for a long while amongst a few razorbills on the low rocky platform of An Toll. Six eggs lay on the ledges, four unguarded and two being indifferently incubated by their owners. One lone chick, new to the world, crouched quiveringly under the cover of a neighbouring boulder; and a pair of oystercatchers beseeched us plaintively to depart, circling endlessly as they uttered their wild cry, sure sign that they had territory on this part with either eggs or chicks. The latter it proved to be, for the Boy, to his delight, suddenly spotted a chick of two or three days—it still had the egg tooth—crouched by a rock, motionless, its downy covering blending with the colouring of the boulders. Later, he had a bigger thrill, for whilst we roved around herring gull territory he excitedly beckoned me to join him.

On the ledge below a vee cleft in the cliff top, only 50 feet above the sea, was a hooded crow's nest with two well-developed chicks. Their tail feathers were just evident and wing primaries were developing. One adult was circling the area, but the chicks remained motionless.

The herring gull colony was a small one: nests with two eggs, nests with three eggs, nests with pipped eggs, and half a dozen downy chicks, aged from two to twelve days, scuttled out of sight as we approached, to hide, motionless, in their favourite rocky crevice. As we left the gulls a seal popped its head out of the water, took a quick, silent look around, and sank swiftly. Near the walled enclosure, by the bothy, two meadow pipits obligingly flew out of the long grass to reveal a nest with three eggs and an empty nest. We marked the spots so that we could check progress—but not too well, as we never again located them.

Rain drew a wet veil over our third day on Handa. How it rained! The island and the space between it and the low cloud blanket was sodden from first light, and an unrelenting downpour continued hour after hour until past dusk. The day was a lost and wet one, and we stayed imprisoned in the bothy, read the visitors' diary, thumbed through the small library of natural-history books, cooked, made innumerable cups of coffee and tea and talked with our fellow islanders.

Mostly the conversation was about birds, but it ranged widely and included the mechanics of sanitation because the local installation was being temperamental. Arthur, who knew something about these matters, diagnosed a stone in the input side, he and Cy nobly performed a delicate operation, and soon personal comfort was restored.

Bare and rude the bothy may appear as it is approached across the meadow; not until you are inside does it reveal that this is no ordinary bothy, but in the five-star class. A noble fireplace, amply proportioned and complete with a pot hook, graces one gabled end. The furnishings and furniture, old-fashioned and discarded pieces, have taken on a new life of usefulness. Many an aching ornithologist has rested weary limbs on that ancient but elegant sofa; that piece would probably fetch its weight in copper if displayed in one of those Kensington shops that sell bric-à-brac —and be as eagerly sought after as the beds that Queen Elizabeth I slept in if it were known that here sat George Waterston, Eric Hosking slept here one stormy night, this muddy stain is believed to have been made by the boot of the Reverend Peter H. T. Hartley or was it Philip Brown?

For the cook and chef there is a kitchen corner with a calor gas cooker and additional boiling rings, pots and pans in plenty; with the greatest of ease a superb banquet can be prepared; from tinned hamburgers, if that is what you've brought, or a Steak Diane if you were imaginative enough to pack this in your tucker box. No duck l'orange, no poultry at all if you please— this *is* a bird sanctuary.

The memories linger yet, and the tastes still tease my palate, of the satisfying meals we prepared late at night in this palace of bothies.

Sleeping accommodation for four or five, perhaps six people is provided in this the living-room. Built on the north side is a dormitory, sleeping six or more. Then comes a touch of genius, the annexe, which is a washroom cum toilet containing a washbasin with a water tap, a water-flushed lavatory and a tap for drawing off water for drinking and cooking needs. The water

required is piped from a small lochan on the hill at the back of
the buildings, and appropriately named the Bothy Loch. The
elevation is high enough to ensure that the head of water is
adequate for all purposes, whether it be the turn of a tap or the
tug of a chain.

When you have tramped half a mile, perhaps further, across
deep heather and bracken, up and down steep hillocks, to fetch
and carry back 2 gallons or more of water (one gallon of water
weighs 10 pounds), maybe in heavy rain or a gale, you will
regard piped water and a tap as the last word in luxury.

Next to this little haven of comfort is the store containing
collected driftwood and dried peats, to which you must add to
ensure continuity of supplies for those who are to come, just as
those who are gone have provided for you. An axe and a saw
are there to reduce driftwood to firewood, a basket for carrying
the peats from the hags and a tushkar for peat cutting.

One of the delightful surprises of this bothy is the store of
materials that it contains, everything that a naturalist, serious or
for a couple of days, may want. There are rope, thread, string,
cord, needle and cotton, fish hooks, lines and canes, clothes pegs,
candles and corks, matches and drying cloths and Tilley lamps—
with spare mantles, paraffin for burning and meths for starting—
to provide illumination in the dark hours.

The bothy buildings have been restored to the present condition
from a near-derelict stone cottage used for many years by the
grazing tenant as a temporary shelter when he came to tend his
sheep. The restoration has been largely due to the efforts, practical
and administrative, of George Waterston and his wife Irene. A
major external improvement was the substitution of asbestos
roofing for the corrugated iron. The battering of rain upon an
iron roof can be very disturbing.

Ruby and Arthur bid us goodbye next morning, under threaten-
ing skies. Alistair Munro picked them up a little after nine, and
in the time it took them to reach Tarbet the Handa skies cleared
to blue. Island weather seems to be quite unpredictable and rarely
does it conform to mainland forecasts. Many times I have sat in
broiling sunshine listening to BBC threats to drown me in rain.
But the pattern can change within minutes. A small black cloud

at sea suddenly envelops the island; it is as night and what was a gentle breeze has worked up to the fury of a gale; the rain comes in horizontal sheets and the whole island is under whip-lash for up to half an hour. As quickly as it approached the storm passes, and cliff and rock and bracken and heather drip diaman-tine under a smiling sun. Even the bigger small islands are fickle, and I know none more capricious than Man. I have been in this nursery of speed on a June morning practising for the greatest of all races, the motor-cycle Tourist Trophies. We would leave Douglas at 4.30 a.m. on our 37¾-m.p.h. lap, enjoy the morning sun as we ran through Kirkmichael to Ramsey and grope our way through thick mist with visibility down to 5 yards as we climbed the Mountain Road.

Taking heart in the clear skies we made for the Great Stack. This was to be a day out, and so it was. From the Stack we worked round the coastline counter-clockwise, and did not return to the bothy until past eleven that night. Except for a sudden and brief shower late in the afternoon, this was a dry day, a day of sunshine with flocks of white, cotton wool clouds chased across the sky by an unruly wind. My! how it blew—in tremendous gusts, at gale force.

As we moved around we saw the havoc created by yesterday's monsoon. Nature treats her children roughly.

Out in the Sound were ten eiders, but only four chicks, and four shelducks without families. The Boy rescued an eider duck-ling from a tangle of seaweed; it was waterlogged, wounded in the neck and struggling feebly. After we had dried it out and brooded it in our woollies it recovered a little so we left it in the shelter of some rocks; but when we returned that way in the evening it was dead.

Not far from the Great Stack, by a deep, water-filled hole, lay a crumpled lamb. There wasn't a mark on it and it had probably died in the night from a combination of exhaustion and exposure. Its young body was not yet stiff, and yet, so soon, the hoodies, or the gulls, had plucked out its eyes. Probably the greater black-backed gulls as there was a small colony nearby. Fierce creatures, the greater black backs. As scavengers and carrion eaters they fill a useful niche, but all too often they don't wait for their victims to die.

Ironically Nature had struck a balance, and a pathetic sight it was. Ten feet ahead of a gull's nest lay the bodies of three young gulls only several days old, in vee formation, legs fully extended behind them, necks and heads stretched out in front; they might have been laid out by an undertaker. Something had struck them down simultaneously but there was not a mark on them. Almost certainly there was no human interference. This was not a spot normally frequented by day visitors—one and all they make for the Great Stack, for that is Handa's showpiece—and no visitors had set foot on the island in yesterday's torrential rains. Until we came along no human had set foot on this territory for at least forty-eight hours. And so the cause of their deaths remained a mystery.

We spent three hours at the Great Stack. This was the Boy's first sight of it, and as my previous visit had been made at a time of heavy rain I was seeing, for the first time, an immense seabird colony—numbering 30,000 or more—at full blast. The scene beggars description, it must be visited to be appreciated to the full. The latter half of June is the ideal time, for the birds have then reached a pitch of maximum activity; eggs are hatching, hungry chicks must be protected and fed.

The Stack is an immense straight-sided block of Torridonian sandstone, tops 350 feet, is roughly 100 feet along each of its four sides and stands on three great feet about which the sea rushes and roars. On a rare, calm day a small boat may explore the cavernous underparts. From top to bottom the cliff walls have stratified horizontally and this weathering has provided countless ledges which the kittiwakes and the auks have not been slow to exploit. The population—infestation might be a better word—is counted in thousands. On the bumps and irregularities of the cliff faces, just out of reach of the waves, the kittiwakes hang their seaweed nests, hundreds of them; then from about mid-height rise gallery upon gallery of guillemot huddles; and whenever the ledges crumble into a jumble of rock are shuffles of razorbills—but their rightful kingdom is at the Stack top, which has collapsed into a shambles of rock. The cap is covered with grass and cushions of sea pinks in which a few score puffins burrow.

The sight is an unforgettable one and the concerted cries of the birds incredibly deafening. As we approached the cliff edge the noise suddenly hit us. So did the smell! As many birds were flying around the Stack as were jostling upon it; the air was full of bundles of feathers, falling, rising, floating, spiralling, whizzing through space in the teeth of a gale-force wind. However strong the wind it never seems to ground the seabirds. Not even the auks, whose stubby little wings whirr and whirr like toy motor-boat propellors; they hang suspended making little or no head-way, are suddenly swept back, come up head into the wind and by dint of sheer determination sweep round the headland. The little puffins, who weigh but a pound, were marvellous to watch. They float up the cliffside to the top, hang like a kite, lose a little way, feet dangling, wings and tail spread, then beat the wind and resolutely forge ahead to their burrows, with four or five fish hanging from their multi-coloured bills. They evidently enjoy such conditions, for in a great merry-go-round they fly for minutes on end, and as some touch down on ledge and cliff top, another takes off.

Reluctantly we moved on. To the immediate west of the Stack is a high cliff, Am Bonair, shaped like the head and beak of some huge bird of prey. The straight sides fall sheer to the sea with hardly a ledge or niche for a bird to rest upon, and at their foot a table of flat rock—used by kittiwakes as a gathering place and a place from where they would suddenly rise in a cloud for no reason at all, "kittiwarking" noisily as they left. The cliff top is covered with dense cushions of sea pinks amongst which nest herring and greater black-backed gulls. Somewhere near seems to be the lapwings' territory. Two pairs were circling on this day and it was around here that we had seen them flying and disputing with the gulls, but never a lapwing chick, egg or nest did we find. The gulls were at several stages in their breeding cycle: eggs, pipped eggs, hatching eggs, newly hatched chicks still damp, and chicks from a few to several days old hiding amongst rocks as the adults screamed their warning cries from aloft.

On and on we went around the crinkly coast, the cliffs now descending as we came down the west side to the south country. After we left the cliffs at Poll Ghlup, smothered with kittiwakes, razorbills and guillemots, a few puffins, some fulmars and a solitary shag guarding three downy youngsters, the birds began to thin out. We looked in at the hoodies' nest, the chicks impassive as ever; found Mary and Cy collecting driftwood at Traigh Shourie—we had parted from them at the Stack hours earlier—and staggered wearily up to the bothy. What a day it had been. We had walked and climbed, stumbled and staggered through peat bog and over rocks, stood and stared at the birds, had our stare returned by the puffins who, alone amongst the wild seafowl, seem to have nothing better to do than emulate human holiday makers. Shags made faces at us, fulmars spat, oystercatchers screamed, guillemots and razorbills nodded in friendly fashion, herring gulls dive-bombed us, the larks sang and the snipe drummed. Fourteen hours after we left it we came back to the bothy, 23.15, and just dusk. If the sun never shone again we had seen Handa at its best, and almost every inch of it.

Next morning we said goodbye to Mary and Cy; they were off to Inverness for a short spell and then making for New Zealand before returning to Canada. Apart from day visitors we now had the isle to ourselves, and on Sunday were in splendid isolation.

Muscle-weary from our previous day's marathon, we spent a lazy day at the pleasant bay of Port an Eilein. The day was warmish, the sky carried a great deal of cloud, and a light breeze blew, occasionally accompanied by something that was wet but was neither mist nor rain.

At high water reefs in this bay are inaccessible, but when the tide has ebbed it gives access to these reefs over an expanse of pebble, stone and rock, the whole smothered with a slippery, slithery carpet of seaweeds.

Whilst we waited for the tide to fall the Boy fished, using lugworm which Alistair Munro had recommended, but without success. I busied myself with cameras, photographing wild flowers which grow in profusion and variety on the grass banks that put a green edge around this secluded little bay, and watching our several visitors. Mostly these were rock pipits and wheatears searching for insects; once a rock pigeon flew across nonstop, but a solitary pied wagtail remained for several hours. Two oystercatchers spent the day shrilly resenting our presence, but, search as I did, not a sign of nest, egg or chick could I find. A cuckoo called to us, some curlew hurried over, and

St Kilda. (*Above*) Eric launches our St Kilda mailboat into the waters of Village Bay. (*Below*) Our leader, Alex Warwick, working on the roof of cottage number one

St Kilda. A village panorama. In the foreground is the army encampment, beyond are ruined cottages and the R.A.F. road winding up the slopes of Mullach Mor, and on the left is the storm beach

occasionally a snipe wheeled around the point. On the edge of the beach, near the grass bank, was a ringed plover's nest containing four eggs—just a sandy hollow decorated with seashell fragments, the colouring of the eggs blending perfectly with the surroundings and rendering them almost invisible. The owners were busy running around the beach but not greatly concerned with incubation duties.

One of the plovers executed a pretty little dance which lasted a few minutes, and repeated parts of it by way of an encore. His mate was as unconcerned about this display as she was of impending motherhood. The dance was performed on a patch of sand and began with bowing movements; then the tail was fanned and bent sharply down to touch the sand. Next the wings were drooped, the feet moved and stamped. Finally the wings were fluttered violently, the bird turned in a full circle, and the left wing was twisted upwards and backwards. The wing movement of the dance was repeated several times with an occasional and sudden jerking of the fanned tail upwards.

As soon as the outgoing tide had drained the bay the Boy explored the reefs from end to end. I joined him in response to excited signals. The reefs stand high above the water even at the spring tides and are beautiful islets. The rocks are covered with bright yellow lichens, and in the sheltered nooks and crannies sufficient soil has collected to support a variety of plant life. Two pairs of herring gulls noisily dive-bombed us for violating their territory, and we found their chicks sheltering in shady hideaways —pretty little creatures of grey fluff, black spotted, only a day or two old, for the egg tooth was still to be seen. The major discovery was two eiders' nests, a few feet apart, one with four and one with five eggs. One was a nervous bird and left hurriedly but the other was a tight sitter. We approached quietly and within her full view; at 20 feet she was unseen so still did she sit and so completely did she merge into the multi-coloured surroundings.

Our hopes for a fine day rose with the barometer next morning. Weatherwise it was a day of disappointments; heavy clouds hid the sun most of the time, and released many showers, light, heavy and torrential. In the morning we honoured the shags in their own bay with our presence. This was deeply resented, not

5

only by the shags, who continuously made faces and barked at
us, but also by the herring gulls who own the rocky slopes above
the bay.

The broad ledges of the cliffs attract the shags to the bay which
takes their name; their large, seaweed nests need ample room.
They are not handsome birds but the iridescent greeny-black
plumage and the lovely jade eyes, offset their ungainly appearance.
In the air they are heavy but strong fliers; in the sea they swim
with ease, the body almost submerged, the neck held almost vertically, and the bill tilted upwards, a very characteristic swimming
pose. On land they are downright awkward. They show their
resentment to human intruders by gaping offensively, shaking
their heads and writhing their long necks; the cocks bark hoarsely
but the hens remain strangely silent.

Theirs is a beautiful little bay. Mostly the cliffs are straight-
sided and in the crevices where soil has lodged seapinks grow.
The narrower ledges are occupied by auks in small numbers; low
down the guillemots crowd together and near the top razorbills,
some cuddling eggs against their white tummies and some with
newly hatched chicks, crouched forlornly and all aquiver. Most
of the shags had well-developed chicks in their nests, some
feathered, others in browny down, and, if anything, were uglier
than their parents.

We spent some time on the slopes above Shag Bay, wandering
about the herring gull colony, affording the gulls dive-bombing
practice. A first experience is alarming, but the gulls are not, in
my experience, dangerous, and can be ignored. They do, however, have a noisome habit of defecating upon your head, and
have a well-timed and accurate aim. Object of the morning's
visit was to record on film the belligerency of shag and gull, and
on tape the equivalent bird expressions of "Homo sapiens go
home".

Torrential showers decided us to go back to the bothy and
eat, abandoning our customary practice of cliff-top picnics. Eating did present a difficulty. Not the mechanics of it, nor the
preparation; having the time to eat was the problem. Finally we
came to terms with our stomachs. Breakfast was taken in the
bothy. For me, the city worker, this was a bowl of cereal; for the
Boy, a growing lad of 14 whose appetite matched his proportions,

it was cereal cum eggs and bacon cum toast (thick) and marma-
lade. During the day, he, well fortified by breakfast, carried a
knapsack containing bread, tinned meat, boiled eggs, cheese,
butter and a flask of coffee. To tempt us were chocolate and
raisins. Surprisingly, we ate little during the day, but at eventide,
when we returned to the bothy, maybe at seven, often eleven, we
had a heartening supper. To save time, because by this hour we
were pleasantly tired and savagely hungry, we evolved what was
to become known as Handa Stew. Into a large pot was poured
water and to this added potatoes, carrots and onions. Packets of
dried soups, tinned spaghetti, baked beans, tomatoes, and corned
beef gave it body. Salt, pepper and other seasonings increased its
piquancy. Always there was a residue, but not enough for two,
so the pot was topped up and Handa Stew became a fixed feast.
On the last night we did eat more than our fill; not from greedi-
ness but to prevent waste.

By late afternoon the weather became temporarily settled,
windy and warm. We made for the southern beaches on a tour
of routine checks and on arrival were shrilly greeted by the noisy
objections of the few pairs of herring gulls living in isolation
from the main colonies, and of the local pair of oystercatchers.
The ringed plovers were still searching the rocks, careless of their
eggs. The eider's nest with five now contained one egg and four
chicks; the second pair of ringed plovers, nesting on adjacent
Traigh Shourie, had added a third egg to their scrape. And the
Boy claimed to have seen a couple of porpoises swimming in
the Sound.

About seven we sat down to our evening repast, keen appetites
sharpened further by the aroma of Handa Stew. The first taste—
always the best—was interrupted in mid-orbit between plate and
mouth by a knock on the bothy door. Visitors. Two. Scarcely
had we done our duties as hosts than more arrived, a small party.
All were curious to look and marvel at the bothy, discuss the
days' events and ask questions whilst they waited for the Tarbet
boatmen to come and take them off.

Most visitors are appreciative of Handa and overwhelmed by
the density of the birds, but not so the male half of our first pair
of arrivals. He was unimpressed and abandoned his intention of
camping on the island, it was far too boggy. Noss, he said, was

far better, the Noup superior to the Stack, and Handa had no
bonxies or gannets. We were aggrieved and sprang to Handa's
defence. We felt we ought to, it was our home. Our visitor had
seen a strange guillemot, Brünnich's, he thought. Silently we
handed the local library copy of the *Field Guide*, and he then
decided that the bird was the common and foolish English
guillemot in winter plumage. Handa certainly was not hot, but
it was not that cold.

At ten the Boy went out for a late evening's fishing, arriving
back seventy-five minutes later at dusk, with a 3-inch fish, which
might have been an immature herring. He had caught several
others of the same size, threw some back and used others as live
bait. This one had greedily swallowed the hook and was promised
as a present to some fortunate razorbill. Later—much later!—it
was presented to a shag, which immediately flew in terror from
its rocky shelf. There are six lochs on Handa but none contain
fish, and so, of course, there are no waterhorses. This is a pity
for all such nature reserves as Handa should have a waterhorse,
to say nothing of a dog. Both black on Handa, for this is Sutherland.

Our seventh day on Handa was a Sunday and so we were alone
because no visitors are brought across on the Sabbath. The
weather pattern followed Saturday's: a strong wind, a wet morning with sudden squalls and torrential downpours of rain; and a
dry, windy, sunny afternoon and evening. We spent an hour
with the shags in their bay to obtain photographs and tape
recordings before moving on to watch that wonderful flying
machine, the fulmar, giving an at-home performance about the
precipitous northern cliffs at Fulmar Bay.

The strong, gusty wind provided ideal conditions for them to
display their mastery in the air. They glided, they sailed, they
swooped, they hovered; occasionally they bent one or both
wings, came up over the cliff edge to peer at us, swept confidently
out to sea, circled back to the cliffs, soared and began another lap,
all accomplished effortlessly. A continual stream of birds was
leaving the cliffs and another returning to mates who were
dutifully incubating their eggs. As often as not the returning bird
was greeted by a long, loud cackle, and there they sat both

clattering at once; it looked like domestic strife, but is part of their display. Despite all this hullaballoo the hen lays but one egg and in the barest of nests.

On one part of the cliffs we were able to scramble down to a nesting fulmar. She (or he) sat on a narrow ledge amongst a beautiful little garden of sea pinks, and was a tight sitter, showing no sign of restlessness until we were within a few feet; then she drew back her head and partly opened her bill menacingly. We edged to within 4 feet of her before she spat, but were ready for the attack; she missed her target—only just—and made no effort to try again. Now that she was without ammunition we moved back a few feet and set up the cameras. From then on she ignored us and was placid and rewarded our patience by obligingly showing us how she turned her egg to ensure its even incubation. This delicate operation is performed with a firm push from the beak and a deft body wiggle. By some equally deft and swift work with a ciné camera we had the entire operation recorded on film.

Lying on the grass growing on this ledge, and 6 feet from the fulmar, was a shag's egg, probably purloined by a hoodie or gull from one of the shags' nests in this area. These shags appeared to be entertaining some cousins, for intermingled with them were six adult and six young cormorants.

In the cushions of thrift and grassland at the back of the cliffs of Fulmar Bay, greater black-backed gulls were nesting. We found one nest containing a newly hatched chick—the down was still damp—and an egg with the beak of the chick visible in the shell gap. Occasionally this unhatched chick cheeped plaintively, whilst the hatched one stumbled about the nest drunkenly, uttered noisy and spirited chirrups and hopefully pecked at the flower heads of nearby thrift. As we had a tape recorder with us on that warm afternoon, the cries of the hatched and unhatched birds were duly recorded, and the antics of the young chick filmed.

We began the homeward trek about nine, returning the way we had come, by way of the Lily Loch, on which the first of the water lilies were opening their cups. A lovely evening, the sun going down slowly, lighting the clouds with a silvery glow—it would not set for another ninety minutes. As we passed the loch a bird flew up from the grass, its white rump clearly visible.

Perhaps she's left her nest, we told each other excitedly, for we hadn't yet found a wheatear's nest. When we reached the point of departure we found, to our surprise, a meadow pipit's nest with four eggs, and marked the spot carefully, for later observation. Before we reached the bothy, we found our only caterpillar, a brown woolly specimen, making its way across the rough pasture land.

The next day was 21st June, the first official day of summer.

Handa celebrated it with a bright morning, blue skies and brilliant sunshine—until noon, then the sky clouded and late in the afternoon rain set in for an hour long spell. The evening was superb; the cool wind dropped and the sun shone until it set.

Fulmar and Puffin Bays were on our itinerary, and there we stopped until the heavy rain sent us squelching back to the bothy, sure that the evening would be washed out. It wasn't; with that sudden surprising change that island weather produces, we were back in sunshine. The gull chick that cheeped from its prison of eggshell yesterday was now a damp chick, nibbling at grass and thrift flower heads. The visiting cormorants were on their rocky perches. And at Puffin Bay the Boy made a new discovery, an oystercatcher's nest out on the end of a grass covered limb of rock, with two eggs. We cleared the wheatear/pipit mystery. In the morning the nest was unattended, one egg was damaged, dented with the white oozing out. On our return journey a meadow pipit flew off the nest, now containing only three eggs— the damaged egg was missing. The wheatear must have been feeding very nearby on our earlier visit.

I spent a lazy evening along the south coast, ending it with a burst of energetic activity, collecting driftwood from Traigh Shourie. Whilst there I looked in on the ringed plover to see if the fourth egg had been laid. The nest had been robbed of all eggs, there were no shell fragments to be found and no sign of the plovers along the shore. The only unexpected finds were two black rabbits, and two snails, one with a yellow and brown shell, the other with a yellow and grey shell, and, excepting the huge black slugs that can be found in most parts of Handa in the evening these were the only specimens of gastrapoda found. As we sat outside the bothy enjoying the last of the day's sunshine, a brown rat scuttled past.

Hopes of a completely dry day had long been abandoned, nevertheless on our ninth day the unexpected happened. The wind had gone round to the north-east, the sun shone from rise to set, the temperature leapt to 70°F and we spent most of the day in shirt sleeves, recklessly abandoning woollies and rain-coats. The sea too had settled, into a long, lazy swell.

To appreciate the full beauty and grandeur of Handa's cliffs they must be seen from the outside. There is no better view point than that offered by a small boat. We had been waiting impatiently for a favourable combination of wind and sea to make the round-the-island voyage possible, and on this perfect summer's day the mission was accomplished. A little boat can get close in to the great cliffs teeming with birds, and there was enough ocean swell to add zest to the voyage; as the boat rose and fell at the whim of the Atlantic, the cliffs crazily reeled to and fro in sympathy. We moved close in to the Stack, jerked our heads backward, straining our neck muscles to the limit looking up the corrugated sides to the top dizzily swaying alarmingly; drifted around the tri-footed base, could almost touch the lower ranks of nesting kittiwakes, who stared down at us unconcernedly. As we approached the smaller stacks and skerries, birds in their hundreds took to the air and water. The guillemots, puffins and razorbills just tumbled in head first, row after row, with military precision; the shags went up like a great fleet of flying boats in ragged formation; and the kittiwakes "kittiwarked" so noisily in concert that the roar of the sea and the pop-pop of our boat engine was drowned by their cries. That was a trip that I wouldn't have missed for all the whisky in Scotland. Never were two ciné cameras more busily employed. A great deal of what we saw was in the camera viewfinder, but the sacrifice was worth it because we can, and often do, relive that breathtaking spectacle of bird-colonized sea cliffs, and as often as not we spot the hitherto unseen antics of some group of birds.

The rest of the day was an anti-climax, spent looking in at old friends—for the last time as we were due to leave in less than forty-eight hours. The eider with her four chicks had put to sea, but the other nest still contained four eggs. The Boy thought he saw a golden eagle over Traigh Shourie being mobbed by six greater black-backed gulls, but not having the glasses with him was unable to get a closer look. The big, brown bird was larger than the gulls, so a golden eagle it probably was. Later, a day visitor, a mature hen, waiting for a boatman to take her off, said she had seen a pair of golden eagles over the centre of the island during the afternoon.

Our last complete day on Handa was as foul as the yesterday was fair. Heavy rain in prolonged spells, a stormy, driving wind, and some sunny periods sums it up.

Many of our evenings in the bothy had been brightened by a blazing fire. On a rain and windswept night we sat snug, enjoying the glowing warmth of peat and the cheerful crackle and flame of salted wood, drowsily making plans for the morrow. These memories stirred our consciences and urged us to the peat hags to cut fuel for the next year's visitors. The hags are a neglected spot; we tidied up the litter, made neat piles of cut peats and cut new ones. In the store house is a tushkar, a Shetland-invented tool that makes peat cutting delightfully simple and speedy; even a novice has no difficulty.

There were a great many things we wanted to do but had not done: finding the snipe's nest near the bothy which a visitor— who departed before we arrived—had reported was in the reed bed 150 yards away and contained four eggs. We never found the nest, instead we discovered a meadow pipit's nest with two chicks, feather quills emerging. We never saw the albino oyster-catcher. We shall never see it now; it was found dead on the beach on 9th July 1967 and was known to have lived on Handa for nineteen years. The remains, if you ever want to see them, are in the Royal Scottish Museum, Edinburgh. There were flower photographs I needed. A visiting botanist had made a very complete survey of the island's flowers and plants, drawn up a map and keyed it in colour. This hung, and probably still does, on the right-hand side of the fireplace in the bothy. On a grey day Handa wears a monotonous green-brown smock of grass and bracken, moss, heather and peat bog, relieved only by the steely glint of the small lochans and tumbling burns, and is edged by stern rock. On a summer's day this sombre mantle is brightly studded with the colours of the flowers of the field, the water meadow and the cliff, by the commonplace, the unexpected and the rare, the shy, the tranquil and the bold. The sunny places

twinkle and the shady nooks are lit with the bright spots, gaily
painted in white and pink and blue and yellow.

Our last evening was a boisterous one. We went up to the
Great Stack to do last-minute filming and recording the vocifer-
ous fowl on tape. Near Swaabie Loch we found brown and white
feathers on the grass and decided that these were from a great
skua; there were persistent rumours that these birds had arrived
on Handa and were nesting. We had found no nest evidence, but
a few minutes later a bonxie did circle overhead, flew lazily out
to sea, circled again, came inland lower down the island, and flew
out of sight. Nearer the Stack lapwings were still disputing with
the gulls. Two hours later rain set in and continued for several
hours. We slid back via Puffin and Fulmar Bays, squished over
the hillside past the Lily and Bothy Lochs and dived into the
bothy, dripping water over the floor like miniature burns. Towards midnight a gale was shrieking across the island, lashing the
rain down on to the bothy roof and turning the burn into a torrent.

After eating—it was the last of Handa Stews—we packed.
When we finally got to bed it was 1.30 tomorrow. The night was
a wild one; the wind and the rain hammered and kept us awake.
Round about five the gale eased a little and the rain turned to
drizzle. By breakfast time the sky had run out of rain, but the
wind was again ferocious. We were to leave at 10.30. Before then
the sky was showing patches of blue. The full havoc caused by
storm we never knew. We did see, sadly, that the ringed plover's
nest on the beach of Port an Eilein was wrecked, the eggs were
scattered and cold. The two birds unconcernedly searched the
rocks for food.

Through the glasses we watched Alistair Munro struggling
against the wind as he crossed the Sound. As we loaded our gear
into his boat he said, "I nearly didn't come for you." We half
wished that he hadn't.

Most visitors to Handa come for the day; others, who come to
stay, have the choice of camping or the bothy. The timidly
adventurous, aching for the thrills of living on a desert island and
the dizzy spectacle of seabird cliff colonies at the height of the
breeding season, but not wanting to forego all bodily comforts,
should choose the bothy; it and Handa are for you.

FOUR

St. Kilda Visited

⸪

One hundred and twenty miles west of Scotland lies a group of storm-racked islands, collectively known as St. Kilda, and owned by the Scottish National Trust. Hirta is the largest, a mere 1575 acres. Four miles to the north-east lies Boreray, and a quarter of a mile to the north-west is Soay. Smallest is Dùn, just a stone's throw away in the south-west.

The St. Kildans left in 1930. Twenty-seven years later a missile-tracking station was built on Hirta, and the army maintains a small detachment in the village meadows. Excepting that, the islands are a nature reserve, administered by the Scottish Nature Conservancy and visited by the fortunate few. They are renowned for their bird colonies, an ancient breed of sheep and a unique sub-specie of mouse.

To many who love small islands and their solitude, St. Kilda is the Ultima Thule.

We, a mixed party of thirteen, arrived in Village Bay at sunrise, in a lively swell. At this hour life is said to be at its lowest ebb. I can vouch for that. Most of the party were sea-sick—not me, vomiting never brings me relief. We had left Mallaig at 1 p.m. the previous day, crossed the Sea of the Hebrides—it was in a passive mood—and had passed through the Sound of Harris at midnight. Four or five hours' straightforward ploughing through the Atlantic remained, so we bid our pilot goodnight, offered a prayer for those in peril on the sea, or kept our fingers crossed, according to our convictions, and turned in.

Our boat, the *Western Isles*, was an ex-fishing vessel of 50 tons, converted for cruising, the property of its skipper, Bruce Watt of Mallaig, and had made the St. Kilda run many times. *Western Isles* was under charter to the Scottish National Trust to take small parties to and from St. Kilda, but as at this time there was a shipping strike we had passengers, mail and goods for the near-lying islands of Eigg, Rum and Canna.

After clearing Canna we were solely a party visiting St. Kilda under the auspices of the Scottish National Trust, which since 1958 has organized parties of volunteer Trust members and takes them to the islands to carry out restoration work on the old village cottages, dry-stone walling and those ancient store houses peculiar to St. Kilda and known as cleits. Our party was the third of the year and under the leadership of Alex Warwick of the Trust. We had with us a very special member, Neil Gillies, a St. Kildan who was born at cottage number eleven in Village Street, about 1896, left the island for the mainland more than fifty years ago and was now making a brief return. The rest of the party were ordinary mortals. There were three women: Kathleen from Cheshire, Olwen up from Sussex and Sylvia from Stirling; Tom was a Wallasey man, Lachie a Scot and Duggie a Yorkshire man; Chris lived in Cheshire; Eric came from Manchester and Alan, 16, from Dumfries. The Boy and I were from Kent.

Our arrival in Village Bay aroused no welcoming halloos from the shore. *Western Isles* tooted and hooted and blared. And nobody cared. So *Western Isles* sailed around the bay, rolled across the bay, tossed in the bay and pitched about the bay—sometimes doing the lot together, in a most alarming manner, to the utter discomfort of most of her passengers—for two hours. And then, belatedly, a dory left the shore and came out to meet us. The occupants very sheepishly explained that they had had a farewell party the night before . . . very late to bed . . . and, er, well you know what it is waking up in the morning after the night before . . . and here they were . . . and would we please come ashore . . . wasn't it a lovely morning, with the sun peeping round the side of Oiseval, and the sea all gold plated . . . quite a swell wasn't there? . . . and had we had a good crossing? So having said our how-de-dos we went ashore, in batches.

Going ashore brought excitements, quite a pantomime it was in those boisterous seas. Because *Western Isles* could not come along-side the jetty we transferred to the dory, which shuttled across

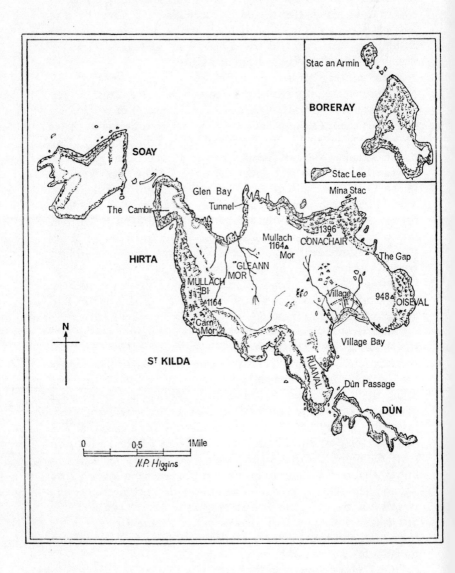

SOAY

BORERAY

Stac an Armin

Stac Lee

Glen Bay

Mina Stac

The Cambir

Tunnel

▲1396

Mullach
1164▲
Mor

CONACHAIR

HIRTA

GLEANN
MOR

The Gap

MULLACH
BI
▲1164

Village

948▲
OISEVAL

Carn
Mor

Village Bay

N

RUAVAL

St KILDA

Dùn Passage

DÙN

0   0·5   1Mile

N.P. Higgins

the half-mile of broken water taking our party off and putting the home-going party aboard. The journey was nothing, in transferring to the dory lay the adventure. The cox was a young man with a powerful voice which he used with little restraint, believing that the greater the clamour the better the result. For the timid his noise was off-putting; timorous hearts and queasy stomachs are not the best of combinations in tumultuous seas. When his turn came, Neil, our St. Kildan, gave the cox a withering look. "Young man", he admonished, "I've done this before," and despite a game leg transferred himself with little more effort than when he went aboard at Mallaig.

Timing was the secret. *Western Isles*, wallowing in a ferment of ocean, rolled and pitched with rhythmic abandon; when she was buried in a trough the dory would come alongside. This wasn't the moment, for she'd suddenly lift her bow over the crest of an on-coming wave, and the dory would slide down and away on the wave's back. Every so often the two boats were together for a few brief moments; in that time one of us stepped over the side and down.

Within about two hours we and our possessions were ashore and the departing crowd were aboard. *Western Isles* circled Village Bay, gave a farewell toot and was away to Mallaig. Wind and sea permitting, we'd see her two weeks hence. Unfavourable weather would see us marooned on Hirta, perhaps for days, extra time to see those sights we were bound to miss. Two weeks is all too short a time. Duggie, who had been to Hirta before, said that on that occasion the party was delayed a week because of bad weather. We got off on time. The most I have ever been delayed by weather is seven hours but, optimistically, I always take extra provisions to keep me going for an additional week.

The Boy was early ashore, he was on deck long before St. Kilda was sighted. Sea-sickness got him there. He came below, on my advice, at midnight, reluctantly, and stayed a short while. The smell of the engine, the stuffiness of the accommodation and the sea building up was too much for a tummy inexperienced in sea-going on small ships and sent him scampering to the rail. Afterwards he took refuge from the cool of the night in one of *Western Isles'* boats. He didn't sleep much, was very wan next morning and hurried ashore at the first opportunity.

A stomach in torment is not a proper companion for sight-seeing from the deck of a small and restless boat in the early morning in Village Bay. So this was St. Kilda, wild, aloof, savage, just as I had read. Before me was a narrow strip of heaving water; beyond, a green shelf; and then a great dark wall of a hill called Conachair that seemed to reach to the sky, oppressive, forbidding, unfriendly, but only because there was no sun upon it—the sun wasn't even high enough to light the ridge. (St. Kilda, ugh! just put me where there is firm ground beneath my feet, then I'll feel human again and appreciate all I see.) The morning was very beautiful. I captured it on film for evermore: *Western Isles* tossing in the bay, the dory shuttling to and fro, provisions coming ashore. Then I went off to the sergeants' mess. The army had a reception committee on duty and were ladling out coffee and tea; the Boy was already there. He had wasted no time and confided that he had found an oystercatcher's nest, with chicks, nearby in a ruined building. A nest with chicks meant that at any moment the family would be legging it for the protection of the meadows, so there was no time to lose; it isn't every day that Nature has film subjects ready and waiting as soon as you step ashore. Very accommodating was that family of oyster-catchers: chicks posing motionless, imitating stones; then, believing themselves to be unobserved, one parent returned to the nest and the chicks hurried back to be brooded. And when that was safely in the can there was time to attend to ordinary matters, such as breakfast, unpacking and settling in.

Amicable relations exist between the Trust and the army. The Trust owns St. Kilda and leases the islands to the Nature Conservancy, who diligently maintains them as a wild-life sanctuary and sub-leases to the army certain areas on the shore and the hill tops of Hirta. The army during its tenancy can carry out all manner of bloody experiments encompassing the destruction of its enemies, but is forbidden to as much as lay a finger on St. Kilda's birds and beasts, not even when the field mice steal chocolate from the soldiers' own pocket. And this command is honoured.

Although we were living on an island returned to nature, wild and remote, life was not primitive. The army has provided the Hirta detachment with many of the comforts that less distant units enjoy. The new St. Kildans generously say to visiting Trust members, "Share with us these good things if you will, we shall enjoy your company." There is the 'Puff Inn', an establishment which caters for most tastes: drinks soft and hard, cigarettes and chocolates, sweetmeats in variety, picture postcards of Village Bay; and, quite unexpectedly, I spied on a shelf a small and familiar yellow box, known the world over and bearing a name synonymous with picture making—Kodak. I have met it in some strange places, but none so remote as St. Kilda. Additionally there is a games room, a cinema, and, to my astonishment, a tellything! The picture received was excellent, but the sound often ran wild. There would be a tantalizing silence, or an incomprehensible babble in some foreign tongue instead of the legitimate noise broadcast by the BBC. There are bodily comforts too: running water, hot and cold; bathroom, shower, wash basins and water-flushed lavatories. Medical help is at hand: we never needed it but in later years a visitor stricken with appendicitis was whizzed back to the mainland and a hospital without delay. Amongst material things, freshly baked bread; and Alex was able to barter foodstuffs which the Queen's men hankered after in exchange for tit-bits we coveted. In that way, one day, there appeared eighteen lamb chops, and it fell to me to cook them.

We were to have slept in large army tents. Shortly before our arrival St. Kilda had had one of its big blows, and before the force of that wind all the tents went down. Instead we were housed in the now-disused church and schoolroom, the women taking the latter. The church was dry and windproof, and very gloomy. As we only slept there the gloom was not of any consequence. I never liked the place, it was unsettling. I am sensitive to buildings. They are friendly or they are unfriendly; given the choice I would have slept elsewhere. My nights were never restless, I slept well. Perhaps in that church still lurked the uneasy presence of a strict sabbatarian missionary, disturbed and agitated because the sacred interior of his church had been violated. I am not a believer in ghosts and spirits, but some buildings have a marked effect upon me, and this old church was disturbing.

Our visit was primarily concerned with restoration work on the fabric of the cottages, rebuilding fallen sections of dry-stone boundary walls and damaged cleits. We worked from nine until one, pausing for elevenses. The rest of the day was spent on our own pleasures. The major jobs were to re-roof cottage number one and rebuild the falling gables of another. At the end of our visit of two weeks cottage number one roof was all but retimbered; following groups completed the timbering and laid on a water-proof covering. Subsequently a second cottage was roofed, and the Trust's visiting parties now use them as a dormitory and living quarters.

The village is not a pretty place. It is stark and empty, lies on a lonely, storm-racked glen, battered and rent by fierce winds. And yet, on a warm summer's day it can be near to paradise; whatever complexion it wears it holds an irresistible fascination for those who love remote islands. Lost within its boundaries are centuries of history, probably millenniums; one day the Boy found a stone axe-head in the street.

The slopes of Oiseval offer a superb viewpoint of the entire village, which is the arena of the huge amphitheatre formed by the high Conachair and Oiseval. Curving across, following the line of the shore, is the street with its sixteen gaunt and ruined cottages, a few standing in collapse; others, stone skeletons, restored so by the Trust. In front of the cottages, meadows slope gently down to the shore. Behind the cottages is a walled grave-yard, and from there the meadows climb to a boundary wall. Scattered at random within this enclosure are several dozen cleits. The street curves down to the shore after leaving the first cottage, past the factor's house, which stands apart, is in good repair and habitable, and on to the school, church and the manse, all three in reasonable repair and standing aloof near the pier.

The cleits—there are several hundred of them scattered across Hirta, in the meadows, on the hillsides and even on the cliffs—are dry-walled store-houses of rough stone, roofed with turf and having a door at one end. They are unique to St. Kilda, a survival from past centuries and still in use. The Trust uses them to store foodstuffs for its visiting parties, and very excellent pantries they make.

We ate our meals out of doors unless it was raining; then we sheltered in a large army tent provided for our use. We did our own cooking and, in pairs, each was cook for a day. I was paired

St Kilda. The southern ramparts of Dùn, gilded with yellow lichen and providing nesting haunts for sea-birds. In the distance is the rock Levenish

A puffin chick, a ball of black fluff that spends the first six weeks of its life in a dark burrow

St Kilda. The field mouse unique to the island

with Kathleen, a school teacher, and confidentially expected to
be her help, an expectation that was quickly and rudely shattered.
"I can't cook," confessed Kathleen, "I hope you can."

And Alex produced eighteen lamb chops, a barter deal with
the army. Our cooking facilities were good—boiling rings fired
by calor gas, pots and pans in plenty, but no oven. I'm not a
lover of the frying pan, and chops cooked in one is an easy way
to ruin good meat. A grilled chop makes a succulent dish; a
baked or roast chop is that much better.

Our roles were now reversed, Kathleen the helper, me the chef.
I told her what I needed, and what I needed most was breadcrumbs. Use corn flakes she suggested, and resolutely refused to
reduce bread to crumbs. Corn flakes it had to be. Meanwhile she
happily peeled a mountain of potatoes, cut up and toasted bread,
and then cut it into small squares. Our main meal for the day was
to be soup with croûtons, roast lamb chops and roast potatoes,
fruit salad. Soup and salad were easy, just a matter of opening
appropriate tins, sloshing the contents of the soup tins into a
pot and heating it; and when the time came for serving, ladling
spoonfuls into the dishes proffered by the hungry mob and
chucking in a handful of croûtons.

Alas, we has no oven. Here we were on a primitive island,
fenced in by a pub, television, cinema, hot and cold running
water, Land-Rovers, electrical generators, radio telephone, radar
and other paraphernalia of modern warfare, and yet we were
without an oven for the roasting of eighteen lamb chops and a
few dozen potatoes.

We managed. Two open baking pans placed one atop the other
and set over a gas burner make an excellent field oven. Keep the
flame low and a watchful eye on the meats, then they won't char,
and out comes a meal fit for a queen. Our success was largely due
to the treatment given to the chops during cooking. What is a
chop? A lot of bone, too much fat and a wee bit red meat; it
needs nourishment. Take several onions, parboil for twenty
minutes, chop small; mix with handfuls of breadcrumbs, add
chopped parsley, sprinkle in salt, pepper—cayenne for preference,
Tabasco is better—ground coriander, cummin and methe, and
wet with milk and water—wine if you can! (we hadn't cayenne,
Tabasco, coriander, cummin and methe, or wine, used powdered

6

milk mixed with water, dried parsley and other herbs). Put cooking oil or fat in the baking pan and heat. Slosh the chops in the hot oil, cover with the stuffing and cook slowly for about an hour.

Our timing was computer perfect. On the instant the mob had spilled the last drop of soup down its several gullets, the chops, roast potatoes and vegetables—tinned I'm afraid!—were ready and served. The meal went down well; greedy fellows came back for more.

After the midday meal, cooks excepted, we were free to go where we wished. The cooks had two or three hours' freedom and then began to prepare evening fodder, taken at about six.

The village meadows provided an infinite variety of interests: flowers for botanists, birds for ornithologists, sheep and mice for mammalogists, old buildings for archaeologists, all these plus wonderful seascapes and views of the island of Dùn across the Bay for photographers and painters, and wide open spaces to look upon and grassy banks to rest upon for idle wanderers. Something for everyone, and no call to walk far or overtax bodily strength.

At the west end of the village was a snipe's nest deeply hidden in a clump of wild flags. The hen was a tight sitter and not at all nervous. We could pass within 8 feet without disturbing her, and at that distance she was quite invisible to a casual passer-by. Knowing that the nest was there we could pick out a shadowy image and see the glint of her eye. She wasn't camera shy. One afternoon I was on cook shift and couldn't wander far afield, I set up a camera, without a hide, gradually inched it nearer, got within 4 feet before the bird decided enough was enough, and off she went to reveal her four beautiful marked eggs.

Although this was the only nest we saw snipe are common breeders on St. Kilda. We could hear them chippering and drumming as we toiled on the cottages, and sometimes see lone birds circling over the meadows; and on our walks about Hirta would often disturb them feeding. At about the hour of dusk they could be heard more than at any other time of the day. Quite an orchestra there was with their drumming, the oystercatchers' piping, the herring gulls' mewing and the deep bass "owk owk" of the greater black backs.

A common sight in the village, and a unique feature of the islands, is the horned Soay sheep, a survival from ancient times; they are mountain goat-like in appearance, fleet and sure of foot, some with dark and some with light-coloured fleece, and as primitive as their ancestors of a thousand years ago. Shepherdless they wander over the hills and come down into the village meadows to graze amongst the cottages and the army's buildings, exhibiting a curious mixture of concern and indifference to human passers-by. Let that passer-by halt, raise a camera or otherwise take an interest in sheep and the animals hurry away. Stalk them and they break into a trot, then a canter, and off goes the entire company, not flocking, but in single file. One of the most difficult feats is to photograph a Soay sheep head on.

The only other land mammal on the islands is the St. Kilda field mouse, a lovely little creature with a reddish-brown back and yellowy underparts. It is different from a mainland field mouse in that the ears are larger, the hind feet bigger and the tail as long as the body, differences which are great enough to warrant separate recognition, and so this pretty little animal is known scientifically as *Apodemus sylvaticus hirtensis*. At almost any time of the day *hirtensis* can be seen scuttling about the army encampment and was a frequent visitor to our field kitchen during the evenings when the cooks had gone.

The Boy set a couple of catch-alive traps in the kitchen on one of their known runs as we were anxious to photograph them. To ensure that the captured animal is comfortable during the night, scraps of paper are pushed into the trap; the mouse tears these into small pieces, builds a nest and settles down happily until released next day. The first two attempts failed, the traps were sprung, and the mice escaped. The third attempt was successful; we caught a fully grown mouse and a smaller specimen; and spent several hours obtaining motion pictures as well as still. *Hirtensis* is an agile creature and when released quickly scampers off to a hiding place. We had to have some control over them and at the same time have them running free, in natural surroundings, and contented. We accomplished this by borrowing from the army several sheets of clear glass, and building an enclosure about 4 feet square. Into this we placed a number of stones, so arranged to provide nooks and crannies, tasties such as oatmeal and chocolate, and introduced the mice. They were quite happy, immediately set about exploring this new place, found and appreciated the food, washed their faces, scampered about the stones and did everything we wanted them to do. We must have trapped the two best-behaved mice on the island.

Ordnance Survey publish a very useful single sheet map of St. Kilda on a scale 6 inches to one mile; to keep this compact, the measurements are 32 inches wide by 23 inches deep, Levenish and Boreray are inset and shown on the same scale. Also shown inset is an enlarged map of the village, on a much greater scale, approximately 25 inches to one mile, and clearly showing the dispositions of the village buildings, the cleits, various enclosures, three buildings of great antiquity and the site of an ancient church.

The most modern of the buildings are the cottages, built in 1860. Alongside them are byres, which were originally houses of the black-house type, built only thirty years earlier. Standing in the meadow some 100 feet in front of the cottages numbers ten and eleven, is an imposing-looking cleit, marked on the map as Lady Grange House. The reason given for the name is that Lady Grange, the wife of the Lord Justice-Clerk of Scotland during the mid-eighteenth century, and something of a nuisance to him and a menace to his affairs, was conveniently removed to far St. Kilda where she spent several years in banishment, and is said to have lived in this building. Very cramped she must have been; people who know a good deal about St. Kilda affairs (Williamson and Boyd) think that this house is a surviving portion of a black house.

More ancient—medieval—is Calum Mor House, standing alongside the burn An t-Sruthan, which has its source at the spring Tobar Childa, and about 70 feet from the boundary wall. Tobar Childa is on the other side of this wall. The house is partly subterranean and takes its name from its builder, Calum Mor, evidently a giant of a man and reputed to have built the house in a day.

A little more than 100 yards south-west of Calum Mor is the site of Christ Church (the building has disappeared during the past 150 years). Adjacent to this is the Earth House, an underground dwelling and believed to be the oldest yet discovered in the village. The other discoveries made, both in 1835, were stone coffins in the meadows, about 70 feet in front of cottage number nine, and an underground chamber down by the present church. The chamber is situated in a small walled enclosure about 200 feet north of the church.

In the vicinity of the Earth and Calum Mor houses a medieval village is believed to have existed. This, the ancient structures mentioned, and another ancient village in the northern part of Hirta, Gleann Mor, are discussed in detail by Williamson and Boyd in their books *St. Kilda Summer* and *Mosaic of Islands*, to which reference should be made by those whose interest has been aroused by my brief account.

These stone buildings and walls provide dry snuggeries for the field mice and for another small creature, the St. Kilda Wren, a bird which has sub-specie status, *Troglodytes troglodytes hirtensis* as it differs in several details from the common *t.t. troglodytes*. The colour is grey-brown on the upper parts instead of russet, paler on the lower parts and with dark brown flecks, the bill is longer and the voice different. There are not a vast number of birds, the estimated population on Hirta being about 100 pairs, up to ten pairs nesting in the village, three, perhaps four, dozen on the western cliffs of Carn Mor and Mullach Bi, two smaller concentrations on Oiseval and Conachair, and the remainder in twos and threes along other parts of the coastline.

For several days a common sight in the village was a young shag being fed by the Boy or Alan, and encouraged to exercise wings and legs; it was a willing pupil and subsequently was returned to the sea. A couple of soldiers had brought the bird to us declaring that they had rescued it from a watery grave, that it was orphaned and in need of care and attention. More than likely the shag had left its nest of its own accord, gone down to the sea without its parents, was looking forlornly derelict after its first wetting and had been 'rescued' by the soldiers.

Hill-climbing was our main leisure-hour activity; it had to be if we wanted to get away from the village. The call of the hills is irresistible. Nearest is Oiseval, a great and green dome rising 900 feet from the Atlantic. Further afield is Conachair. At 1,397 feet it is Britain's highest sea cliff. Sometimes we went walking as a party, led by Alex; these were intensive foot-slogging jaunts which took us around, over and up-and-down the hills and glens, left us with broad and never-to-be-forgotten impressions of St. Kilda's grandeur but lacked the intimate glimpses of the awe-inspiring cliffs that only idle wandering and clambering reveals. A sure way of finding these breathtaking scenes is to follow at random, and with no great haste, the sheep paths. These wind around dizzy headlands from which the view is never less than spectacular. Below, is the sapphire sea; above, the azure sky; westward lies Dùn; and 4 miles away to the east the hazy mass of Boreray and its dominant stacks, Armin and Lee.

From Oiseval's top we could look westward across the breath-taking panorama of Village Bay to the tremendous rock fall at Ruaival. From the village this is an easy walk across springy turf, a pleasant saunter along the edge of the low cliffs on the earthy slopes of which grow buttercups, primroses and orchis in profusion. The rocky platform 50 to 80 feet below, about which boisterous seas splash and gurgle, was fringed with shags; in the middle of this wilderness is a small oasis, a wee bit rock pool and tufts of sea pinks, and there sat, as still and silent as the rocks themselves, an eider incubating her eggs. The walk ends with a choice of a scramble up to Ruaival's windy summit, or down to sea level and the narrow Dùn Passage where kittiwakes nest on the cliff face.

Dùn is less than 100 yards away. At ebb tide boulders are uncovered and it is possible to jump from one to another and so reach the foot of the Dùn cliffs, which at their lowest level are some 100 feet high. A good climber can lift himself to the top at one point, and open up all Dùn.

On a sunny and warm evening we would stop on the grassy slopes by Geo Leibli, a near 100-foot cliff on which kittiwakes hang their nests, and gaze idly across the narrow channel. There was always plenty to watch: a lively sea in Village Bay breaking over the rocks and rushing madly into the eastern entrance of the channel, and on the western side, water, smooth and glassy; occasionally a seal head floated by, sank and was gone for ever; over Dùn's hilltops midge-like puffins buzzed in their hundreds; eiders sailed the waters unhurriedly. I've never seen an eider duck flustered by troubled seas; however rough the waters the bird floats over them unconcernedly, everything under control, rise up the front of a wave, sit on the crest, slide down the back, never a feather out of place, never a lack of poise, no panic, quite unruffled, elder-statesmen of the seas.

When we tired of these sports we climbed to Ruaival's peak, scrambling up the tumble of boulders or walking round the rock pile and then doubling back up a steep and stony path that skirts a 400-foot sea cliff and passes beneath a remarkable natural arch, the 'Mistress Stone', which is a great upstanding pillar of rock supporting a massive horizontal slab. By repute this was a place where young St. Kildans performed dizzy balancing feats when competing for the hand of some village Helen.

Amongst the boulder pile on Ruaival's crown are niches just large enough for a man, camera and tripod, overlooking the chasm that divides Hirta from Dùn. These are fine places to be in when the wind is at gale force, blowing the sea fowl hither and thither, excepting, of course, the fulmars, who make light of the fiercest gale. Not so the puffins; the poor valiant puffins, how they suffered, fighting the wind all the way from the Bay, across Dùn and through the Passage, and then when they came to the Ruaival corner and lost the shelter of the cliffs the cross blast picked them up and threw them back to Dùn's hilltops. On they came again, tails fanned, feet dangling, wings spread, rigid, fighting every inch of the way until they gained the protection of the cliffs. Sometimes they did, more often they didn't; they couldn't give up, they had to keep trying, for down in murky burrows rolled-up balls of black, downy, fluffy, puffin chicks were waiting for their suppers.

A rough mile from the 'Mistress Stone' is another show-off piece, the 'Lover's Stone', a large slab of super-elevated rock. The feat is not in the climbing but balancing on the heels on the far edge, then bending down to touch your toes; a not impossible feat for those who possess good balance, as some bold spirits demonstrated.

The walk to the 'Lover's Stone' from Ruaival could be a dull slog over moorland, up the slopes to Mullach Sgar and across to the Stone. A much more interesting route is to follow the sea-cliff line, crossing Na h-Eagen, over or around a small burn, Amhuinn Gleshgil, and then along the cliff heads of Leathaid a Sgithoil Chaoil which lie 500 feet below Claigeann Mor. Finally there is a stiff, heart-thumping climb to Claigeann Mor itself, and the 'Lover's Stone' is gained. This is a slow route because the cliffs and the birds are for ever claiming attention, and it is recommended for those who are easily puffed; every few yards there is an excuse to pause and admire the cliff scenery and at the same time regain breath.

These cliffs are ragged-topped, falling and rising steeply, sometimes dropping sheer, and in places tenanted by fulmars. At the brinks are little platforms and ledges, easy of access, some conveniently protected by rock barricades behind which we could crouch and wedge ourselves and so peep safely over the cliffside to watch the fulmars on their nests. They would crane their necks upwards and regard us with equal curiosity.

The Boy and I spent many happy hours on cliff ledges, suspended between sky and sea, watching fulmars sail and puffins whirr, and the seas crashing against the rocks. Sometimes clouds would descend and hide the hilltops; sometimes a sea mist would creep in, envelop the entire headland and fog the sun. Rock and bird would appear pale and unreal in a ghostly blue light as sun and mist battled for supremacy. When the sun conquered a cloth of gold spread across the islands again. If the mist and clouds triumphed there was neither heaven nor earth; time to go, with wary step, back to our gloomy dormitory, there to exchange mist-sodden clothing for dry garments, and then seek something warming—to ward off chills!—at the 'Puff Inn'.

Another inviting walk, shorter than that to Ruaival but much more strenuous, is the half-mile ascent to the sea cliffs at the back of the village. This delectable spot is known as The Gap; here the slopes of Conachair and Oiseval meet at a common level, more than 500 feet above the sea. The severity of the climb is softened by following the so-called Dry Burn—Amhuinn Ilishgill properly—to its source in the hollow of An Lag Bho'n Tuath, as the wayside distractions of wild flowers and cleits offer constant excuses for making discreet halts and so regain breath. The backward views of the sweeping Bay and the Dùn heights are worth more than a passing glance, for the scene changes perceptibly as each slightly higher elevation is reached; so, too, do the views of Conachair and Oiseval on either side. By the time the head of the burn is gained half the distance has been covered. The beckoning hilltop inspires the climber to ignore aching muscles and make a final spurt, and if this last effort does not take the breath away completely, the first sight of The Gap will assuredly do so. The view is heady, for the cliffs drop dizzily to the sea; on the ledges sit incubating fulmars, their mates afloat in space, borne stiff-winged on the ever-present wind that haunts this playground of bird and sea. This is a place to be in when the wind is roughly playful and comes gusting over the clifftop, and curious fulmars and puffins ride up to peer questioningly at *homo sapiens et puer* sitting there with cameras or lying flat on their tummies, heads over the rocky edge, watching the sea crashing against the cliffs and surging noisily into unseen caves. And when that amusement palls there is always distant Boreray and the whitened stacks, Armin and Lee, to distract the gaze; disturbing too because an insatiable longing arises—to be upon them looking across to Hirta.

In all our hours amongst Hirta's hills and glens an interminable tug-of-war raged within us: the reluctance to tear away from the savage beauty immediately about us opposed by the undeniable facts that there are many other things to see and that our time was limited. Better a brief acquaintance than none at all; better a moment only of satisfaction than to be eternally nagged by an empty frame in memory's picture gallery. So always, eventually, we plodded on. Plod it must be; in hurry too much is missed. There is more to St. Kilda than wild seascapes and mighty hills, splendid though they be. Had we marched on relentlessly, intent only on circumscribing the island, hopping, as it were, from hill-top to hilltop, glen over glen, we would have missed many a beautiful cameo that our cameras recorded for all time: a fulmar stopped short in regaining her egg as we unexpectedly plunged into her apartment secreted in an inland garden rockery; a nest of newly hatched chicks of the greater black-backed gull, chirruping plaintively on a cold and windy hillside; sombre twites that sang to us from their perch on a rock only a few feet away from where we stretched luxuriantly in the sun, lazing on a primrose-strewn bank; an incubating eider, cunningly hidden, only her head protruding above the flowering thrift and waving grasses; sudden encounters with alarmed and leggy gull chicks and juvenile oystercatchers. At almost every turn there was some new excitement and surprise.

So, with reluctance, we left The Gap and plodded upwards to Conachair's peak, following the cliff edge and watching, with never ceasing wonder, countless fulmars, free in their natural element, the turbulent air, or cliff-bound by incubation duties. Nature has especially sculptured St. Kilda's cliffs to accommodate fulmars. Almost every cliff has its colony, small or large, and with such an expanse of rock and choice of nesting sites the colonies are not overcrowded. The great surprise is that there are so many birds present. A population on Hirta of 20,000 pairs has been estimated, and of these 6,000 pairs have colonized the Conachair heights. An impressive feature of this cliff is the great headland Ard Uachdarachd; alongside it, in the sea, is isolated Mina Stac, 212 feet high, but, viewed from a windy brink nearly 1,400 feet above, looking nothing more than a fallen boulder.

Conachair's cliff looks impregnable until ant-size sheep transform wistful gaze into unbelieving stare. Where sheep go man may follow, not necessarily in safety. When unperceived obstacles unexpectedly appear I sometimes wish that I had been less sanguine. The silent evidence of bleached bones, or a smelly corpse rent and torn by the vicious bills of hoodie crow and black-backed gull, and lying on the rocks below, emphasizes that even a nimble Soay sheep can miss its foothold.

The breathtaking panorama of Conachair's seaboard does not exhaust Hirta's store of wonders; there are more to come. Westward lies a new world, inviting exploration. The green slopes of Glacan Mor and Mullach Mor sweep down to Glen Bay, and then more hills climb sharply to 1,000 feet. At the foot of Glacan Mor lies a high coastal flat, a green shelf 200 feet above the sea. When we were on Hirta this was an inhabited spot, the site of a small oceanic research station, the staff provisioned and relieved at regular intervals by helicopter from a naval vessel. Below this verdant pasture is a tunnel, one entrance protected from the open sea by the rocky arm of Gob na h-Airde, the other in Glen Bay and reached by a natural path along the cliff. The entrances are festooned with a huddle of guillemots and nesting kittiwakes. A broad, sloping and slippery platform gives access to the seaward opening in the east through which can be seen distant Boreray and its attendant stacks Armin and Lee. Overhead is a roof of solid rock, rudely carved and unexpectedly colourful. The narrow channel between platform and tunnel wall is a seal's play-ground.

We spent several happy hours in and about the tunnel—once a party of turnstones flew in and entertained us. It is a place of endless fascination and seen at its best on a wild day when the elemental power of the wind is unleashed and spent in howling fury upon the sea, whipping it into boiling foam amongst the rock cisterns.

Glen Bay is deep and lonely. Rarely do ships seek its waters, and landing is a chancy business. The western side of the bay is dominated by a 700-foot hump of a cliff, The Cambir. From the lofty summit we looked out over the sound to Soay which climbs 500 feet higher. Four hundred feet below are the tops of Stac Soay and Stac Biorach, the latter crowned with a guillemot colony. Southward is the soaring cliff of Mullach Bi, and from there the hill ridge reaches out eastward to the summit of Mullach Mor, with Conachair peeping over the rim—and then the whole island falls abruptly to the sea.

The cliffs of The Cambir are steep and wild, colonized by 1,000 fulmars. We spent an entire afternoon sat amongst rocky pin-nacles watching them. These birds, which are such clumsy, sluggish creatures when earthbound, show a lazy gracefulness when gliding around a bay on a windless day, but when the air is violently disturbed they attune themselves to the fury of the storm—wind never grounds a fulmar. On this afternoon a gale was blowing in from the Atlantic so we were treated to the most spectacular flying displays we have ever seen. There is no doubt at all that the fulmar is the perfect flying machine; the mastery they exhibited held us spellbound, and to our great delight part of the performance was given only feet away from where we sat, nursing cameras, patiently waiting for such a moment. One bird posed for me superbly, wings spread and loose not board-stiff, the whole bird alive and trembling, tail spread, legs dangling loosely as an aid to balancing, hanging quiveringly in space, no longer a still-life bird, a painted bird. It dropped slowly, inch by inch. Suddenly a joyous sweep, away across the Sound, lifting up again to Soay's tops and then in a mighty sweeping arc back to The Cambir and another catch-as-catch can game with the wind.

Oh for the wings of a dove.
Far away would I fly.

Mendelssohn could never have seen a fulmar.

There is a choice of a high road and a low road back to the
village, both of never-ending interest. The high road follows the
hill ridges which climb from 400 feet at the foot of The Cambir to
nearly 1,200 feet at the peak of Mullach Bi. The low road is through
Gleann Mor following the burn Amhuinn a Ghlinne Mhoir.

The ridgway is the more tempting, and time consuming,
especially when the temptation to explore sheep paths is not
resisted, for these lead to all manner of exciting prospects which
need further investigation. The Glen has its own attractions: a
small herring gull colony, a larger greater black-backed gull
colony and the remains of an ancient village. Bonxies were
suspected to be present and nesting. We found no nests. We did
see some large birds which might have been bonxies, but they
never came near enough to be positively identified.

The upper slopes of Gleann Mor are strewn with the wreckage
of a Sunderland flying boat which crashed on Hirta during the
war years of 1939 to 1945. The impact when the aircraft hit the
hillside must have been fearful. Huge pieces of the wings and
fuselage lie on the hillside, and three of the great engines are
almost complete, partially buried in the earth.

As we came up the Glen on one occasion a pair of oystercatchers screamed their heads off and buzzed around us like
angry bees, so we halted to watch the birds, and carelessly leant
on a large piece of mouldering aluminium that had once been a
Sunderland wing. The reason for their outburst lay crouching in
the wreckage of the aircraft wing, stone still—a juvenile oystercatcher, trying to pretend it wasn't there. The Boy picked up the
bird and it lay patiently still in his hands, but the moment he set
it on the ground it ran headlong for cover—ran, not flew.

If noise is the yardstick by which the number of birds present
is assessed there must be more oystercatchers on Hirta than all
the other birds put together. Their strident calls are incessant,
echoing across the hills from soon after dawn until past midnight.
The population is not a large one; we never saw great flocks of
the birds. There were usually four to six, a dozen was unusual.

We did encounter them in many places, in the glens, on the hills and by the shore. About fifty to sixty pairs is the number estimated by Williamson and Boyd.

The oystercatcher's call is wild and plaintive, the voice of the moorland and the lonely seashore, a lament. There is another mood, a piping song which begins with the shrill "kleep kleep", builds up into a rapid repetition to become a long trill and is performed by several birds together. Only once I have seen them. Not in St. Kilda, but on the Treshnish Isles, when, one evening, I came upon a flock of some twenty birds on the shore, packed into a tight circle, heads pointed inwards and downwards. My appearance broke the magic circle. In a moment they were gone, "pic picing" agitatedly. I thought I could never tire of their calls, but on St. Kilda they become wearisome. From morn till night, day after day.

There were other songs to listen to: the thin piping of twites and pipits, the warbling of wheatears, the twitter and chatter of starlings; and there was a dawn chorus, led by the wrens. The starlings moved in a flock of up to forty birds, adult and juvenile together. We found no nests. Most likely the stone walls and the cleits are their nesting haunts. The starlings the Boy and I saw were up on the Ruaival and Mullach Sgar heights; there was none in the village. We saw only one flock at any time, and this was probably Hirta's entire population. Nor did we see any starling moots in the mornings or evenings. There was no need for them, it didn't take long to fly from one side of Hirta to another. The birds that did circulate in great aerial flocks of an evening, and almost until midnight, were the puffins. There was always a great cloud of them over the large puffinries on Dùn and at Carn Mor on Hirta.

Carn Mor is a place of special interest, a place that must be visited—not at a sensible hour, but after midnight, for this is the home of a nocturnal and rarely seen bird. Leach's fork-tailed petrel, a bird of the ocean that comes ashore only to breed, nesting in an earthy burrow or under stones. There are only four known breeding stations in the British Isles and St. Kilda is one of them.

The shortest way to Carn Mor from the village is straight up the side of Mullach Sgar, an 800-foot climb in half a mile. There

is an easier way, along the road made by the R.A.F. task force in
1957. This zig-zags up the hillside, alpine road fashion, leading
to the radar scanners on the hill tops. That was the route the Boy
and I followed, because we were well-laden: tape recorder,
cameras, food and drink and a gas-fired (calor) lantern. I wanted
this for illumination whilst taking night-time photographs. I
cursed the thing thoroughly all the way to Carn Mor, especially
when scrambling amongst the boulders; and more so on the
return journey when I was tired and sleepy. The lamp was never
meant for such a purpose; it was made to light a tent or cottage.
The illumination was very good, the construction was fragile;
that is, there was the usual incandescent mantle on the burner,
surrounded by a glass wind shield. Handle carefully, very
fragile. I bought the lamp especially for this purpose, and it was
successful. Unfortunately we never managed to locate any 'fork-
tails' in their burrows. Nevertheless we did get some ciné shots
of ourselves drinking coffee, of the cliffs, the sea and gulls, on
black and white film, speed 27 D.I.N.

We had set off fairly early and once on top of the hills walked
along the old wall that overlooks Gleann Mor. Just before eleven
we began the descent on to Carn Mor. An hour to a summer
midnight, the moon slowly sinking. A keen wind came off the
Atlantic, hissed through the boulders and turned up our coat
collars. On the shore, 500 feet below, the sea sucked and roared
monotonously about the rocks. A small colony of gulls was
making a noisy chorus. Unexpectedly an oystercatcher broke
cover, circled the headland, its strident cry echoing off the hillside, rising to a crescendo and dying away plaintively. Distantly
a snipe was drumming.

The cliffside over which we were walking was a steep, boulder-
strewn terrace; little boulders, large boulders, boulders the size
of a house. Beneath lived puffins by the thousand. Some were
already abed, lowing like cattle; others were still flying, taking
part in the great fly-past that was a regular evening feature. Not
all puffinries indulge in these aerial displays; I once lived with a
small colony and they were very sober birds, putting themselves
to bed by 10.30 regularly.

At intervals the boulder field was interrupted by grass-grown
spaces into which puffins and petrels can burrow as an alternative

to nesting in the caverns of the boulder jungle. We halted on the edge of one of these, crouched amongst boulders to avoid the wind, and settled down to wait.

An hour later the moon had gone. For a long time all was fairly quiet. The gulls were silent, the puffins had retired, only occasionally did we hear their lowing; there was just the ceaseless wash of the sea and the moaning of the wind. Then, at last, came the moment we were awaiting. Strange dark shapes began to flit about our heads, and soft, querulous chuckling noises filled the air, died away, came again, and were gone. The 'fork-tails' were coming in from the sea. A short wait and another wave fluttered around us. Occasionally we could hear a bird crooning weirdly in its burrow. We searched but could not find a nesting site. Once there was a 'whoosh' as something dark and large swept past our heads, and a wild strange cry startled us—rather like an express train shrieking from a tunnel, but nothing more than a shearwater leaving its burrow and throwing itself off the cliff. The Hebridean night in June is a short one. A little after three the eastern sky begins to lighten, and the seabird nocturne ends. Slowly and wearily we clambered back over the boulder field, laboriously climbed to the peak of Mullach Bi to see a new day being born. Away to the west is night, to the east the glow from the not yet risen sun lights the rim of distant hills. As we followed the hill ridge we could see far off islands all aglow. We stood suspended between night and day. Overhead a greater black-backed gull circled endlessly, croaking monotonously, as it searched for its breakfast; and up went a pair of oystercatchers screaming noisly. Quickly the light strengthened. Sheep stretched lazily and moved off to munch another day away; and above the murmur of the sea 1,000 feet below, rose the morning song of the wrens.

Wearily, then, we made our way back to camp to snatch an hour or so of sleep. I had to be astir early; I was on cook's duty. The cookhouse was no show place. A shack, protected from the weather but permanently open on one side, looking out on to an open-air scullery, where we washed cutlery and dishes and scoured cooking pots and pans. The scullery commanded an excellent all-round view—the hills, the village, the bay and Dùn —and so relieved the monotony of eyefuls of potatoes, pots of soup and other fodder. In front of the cookhouse was our open-air dining room, pleasantly green and adequately provided with seating accommodation: a wooden form, large stones of convenient height and contour. Higher up the slope herring gulls gathered, patiently waiting and vigorously fighting for scraps. This opened the way for easy photography. With camera, tripod mounted and focused on a particular spot, tit-bits were then strewn on the grass; along came the gulls; press the shutter release; birds photographed while they waited. In practice there were difficulties, the birds would cheat by nipping in, on the wing, grabbing a choice morsel and flying off. Guile was met by guile. The bait was pinned to the ground. That foxed them.

The view of Village Bay was excellent and there was always something to watch out there, if only the sea dashing itself into foam against Dùn's stony edge. Occasionally there was the excitement of visitors. The *Mull* came regularly on military affairs; more romantic were Spanish trawlers.

The weather was mixed. Island weather, of course. There were days of almost tropical heat, temperatures in the low seventies but so intensely hot that bare skin was burned. Then the wind would change to another quarter. Squalls, rain and high wind forecasted Neil Gillies. He was right too! Some days were dark, some just dull; one was all wet; wet starters would change to brilliant, windy, sunny afternoons, and as often as not revert to wetness in the evenings; and sunny days deteriorated into drippers. Just the usual pattern of island weather, almost unpredictable, except to those born to the job, such as Neil—uncannily right he was.

Not only did Neil know his native St. Kilda weather, he also knew all about mailboats, and made one for us. In the far off days the St. Kildans used them as an ocean mail to get messages to the mainland. The boat is a block of wood, shaped to a point at one end, and the centre hollowed out. A message would be put into a bottle or tin can, which was sealed with candle grease to keep out water, and the bottle put inside the boat's hull; then a lid was nailed on, marked "please open" in poker-work, the boat fitted with a 'mast', and the whole contraption fastened to a sheep's bladder. We hadn't a bladder so used a spherical float that was lying on the shore, probably off a lobster pot. We all put inside a

Shiant. The cottage on the shore of Eilean an Tighe

Shiant. (*Above*) Looking westward from the cottage on Eilean an Tighe. It is midnight and all is calm. (*Below*) The same view by day when the wind had put the sea into a great commotion

card or envelope addressed to ourselves and stamped, and Neil Gillies signed his name. In with the envelopes we put silver coins as a reward to the finder, asking him to post the cards and envelopes. The launching was quite a ceremony. We all went down to the shore, a little east of the manse, Eric took the mail boat and flung it into the sea. and the sea shortly afterwards tossed it landward. The boat was hoplessly tangled in seaweed, so Chris donned swimming trunks, retrieved the boat and swam with it out into the bay; from there a favourable current bore it far out into the mighty ocean. At about 10.30 a.m. on Thursday 30th June 1966 yet another in the long line of St. Kilda mailboats was successfully launched. About three weeks later a postman pushed two envelopes through our letter box, one addressed to the Boy, the other to me. They were postmarked 15th July 10 p.m. Stornaway, so had taken about two weeks to sail across the seas to Lewis. Water got into the container, the ink was waterproof but the gum sealing the envelope's flaps had unstuck; the sender sealed them with Sellotape.

The launching day was a dull one. The Boy and I walked over to the tunnel at Gob na h-Airde in the afternoon. We spent a long time there. The seas were wild and tempestuous and the clouds were low—no day for photography, we hadn't any artificial light. We had intended to spend part of the time on The Cambir but that was in cloud; instead we wandered about the ancient village in Gleann Mor, followed the burn Amhuinn a' Ghlinne Mhoir up to its source on Mullach Bi and examined the various pieces of wrecked Sunderland flying boat, trying to figure out what part of the craft they had been. The clouds dropped lower and lower and, as we followed the old wall, enveloped us. Visibility was only a few yards; the silence was as heavy as the mist. The wall was an excellent landmark and just before we reached its end on Mullach Geal, the mist turned to rain. To shorten—we hoped—the journey, we abandoned our original plan to walk down the R.A.F. roadway, for visibility was now greatly improved. We slithered down the hillside and were thoroughly soaked about the legs by the wet vegetation. The road would have been better. The village was ever distant. After what seemed an age we crossed Amhuin Mor by the Bailey bridge and squelched along the shore, back to the gloomy

7

church, our spirits just as dark. Dusk was falling, 11 p.m. Most of the company were abed. We followed quickly.

Next morning was wet, and so was the afternoon and evening. The sky looked as though it could never be bright again. Next morning it was all we could wish for, and Dùn stood out sharp and clear against the sky. The other islands are a continuous torment. A landing depends on a rare and favourable combination of wind and sea, and mostly the seas are boisterous. We were lucky; a day of calm waters and gentle swell enabled all of us to tumble on to Dùn. As we crossed the bay the sea was full of fishing puffins, so gorged that they could not rise, only beat furiously across the surface as our dory came amongst them.

Landing had its excitements. There was no beach, and the small boat rose and fell on the swell. We simply leapt on to the slippery rocks, and first ashore lent helping hands to those who followed.

Dùn is saturated with birds. All the usual colonists are there: fulmars, gulls, guillemots, puffins, razorbills and shags; flying, swimming, pottering, idling, sunning, preening, then, suddenly aghast at our unexpected invasion, noisily taking to the air. Down in the rocky crevices parents squawked alarms to their chicks.

Boreray and the stacks Armin and Lee are an oft-recurring sight. Here is Europe's biggest gannetry, some 40,000 pairs of birds nesting on stacks and island, transforming the dark rock to a dazzling white. At 627 and 544 feet Armin and Lee are the tallest stacks in Britain. But they remained for us a distant prospect.

Alex alone managed a landing on Soay. It happened this way. The oceanic research station at Gob na h-Airde was relieved and provisioned by a helicopter flown from off a naval boat. During its sorties from boat to shore, bringing in foodstuffs and equipment and exchanging the men on duty, it whisked Alex away to Soay's 1,200-foot summit, and left him there to enjoy seventy minutes' glorious exploration.

Early next morning we left St. Kilda. A ship's siren echoed across Village Bay, bouncing off Oiseval, Conachair and Dùn's crags, and hauled us out of sleeping bags in the small, dark hours. False alarm. Just a military craft on army business. Hours later, at the time when the sun dispels the night, there was another call. *Western Isles* was in the bay. She had had a restless journey. The ship-to-shore panto began again; the visiting party exchanged places with us. *Western Isles* circled the bay, gave a farewell toot. And we left as we came, in a cold light and a heavy swell, with a grey mist on the face of the sea.

FIVE

A Cottage on the Shore

·⚓·

On a low coastal flat of a small Hebridean island stands a little stone cottage, empty, deserted, alone. We came to this enchanted place on a golden afternoon in June, and the master of our boat put us down on a stony beach, promising to retrieve us two weeks hence.

The mapmaker calls these isles the Shiants. In the musical tongue of the Gaels they are known as Eileanan Seunta, which means, the Enchanted Isles. There are three. Eilean an Tighe (the Island of the House) is joined to Garbh Eilean (the Rough Island) by the stony beach, and 400 yards across the bay lies Eilean Mhuire (Mary Island).

For two weeks these islands were our playground and the cottage was home.

There we were, the Boy and I, on a sunny afternoon in June, sitting in our car writing holiday postcards—"having a wonderful time . . ."—opposite the post office in Tarbert, Harris, when up comes a native of these parts and asks in the soft accent of the Islanders, "Are you the party for the Shiants?"

You could have knocked me down with a postage stamp. We were in a place we have never before visited, 800 miles from our home in Kent, knowing no one, known by none. Our destination was a group of three small and uninhabited islands lying in the Minch some 20 miles from Tarbert. We had been in Tarbert a bare fifteen minutes—and were discovered!

I regarded my questioner with amazement and puzzled suspicion; how did he know about us and our movements? Grudgingly I admitted that we were bound for the Shiants. Then swiftly came the explanation. I had hired a fishing boat to take us on the last leg of our journey, and the owner, Roderick Cunningham, lives on the small island of Scalpay, which lies inshore by Tarbert. We were to have joined him on Scalpay in the evening, but he, having business in Tarbert that morning, had kindly decided to meet us off the Hebridean Ferry from Skye, so saving us several hours. A message had gone out over the public address system of the ferry boat, and this we had missed.

Scramble! We were back at the jetty without delay, unloaded our pile of goods into the boat, the *Isle of Skye*, dumped the car in the nearby public car park, and took ourselves aboard. Within the hour we had put to sea and were off the Shiants in about two hours more, sailing close under great cliffs that rise a full 400 feet from the blue waters of the Minch. The day was summer perfection, the sea a glass mirror, the thermometer was in the seventies, and the lightest of breezes gently fanned us.

Going ashore was a leisurely affair. *Isle of Skye* anchored in the bay and put out her boat. The Boy dropped easily into it, our goods followed, then me. A gentle row to a pebbly beach, the bar which joins an Tighe to Garbh; step ashore, unload the boat. Within minutes it was being lifted back on to *Isle of Skye*. None of the wild, dawn pantomime that faces the visitor to far-off St. Kilda.

We stood on the landing beach amidst our scattered belongings watching our boat glide across the bay and through the strait that separates Garbh from Mhuire, watching until her stern disappeared behind the cliffs. Now we were alone. Around us lay a score and more of packages waiting to be carted over a rough quarter of a mile to a cottage. We had brought with us everything we would need: cartons of food and drink, cases of clothes and bedding, photographic and tape-recording gear, fishing rods and line, miscellaneous camping gear, for these islands lie many miles from human habitation and have long been deserted save for occasional visitors such as we.

Within an hour our portering was done. Two by two we humped the packages across the stone beach, climbed a crazy rock staircase to a coastal flat on the west shore of an Tighe, ambled carelessly through a meadow bright with summer flowers, and halted, a trifle puffed, by a wee stone cot, its walls shimmering

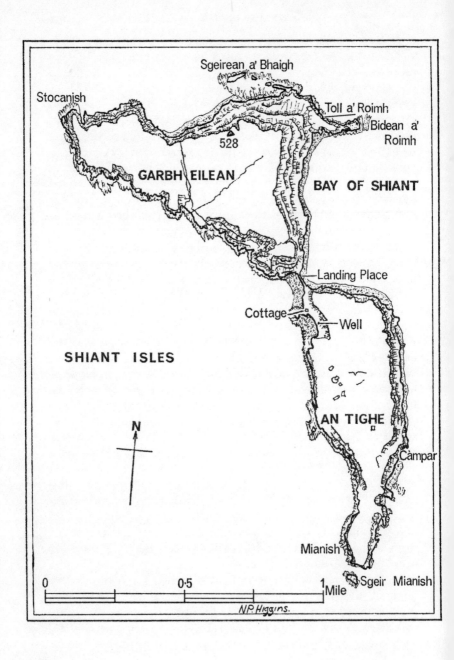

Sgeirean a' Bhaigh

Stocanish

Toll a' Roimh

Bidean a' Roimh

528

GARBH EILEAN

BAY OF SHIANT

Landing Place

Cottage

Well

SHIANT ISLES

AN TIGHE

Campar

N

Mianish

Sgeir Mianish

0   0·5   1
                              Mile

N.P. Higgins.

white in the glare of a burning sun. Tedious work, especially on such a summer's day; doubly irksome on a day of wind and lashing rain. Over exertion is avoided by keeping the packages as light as possible, 15 to 20 pounds apiece; many of light weight are preferable to a few, heavy and cumbersome.

The cottage stank. We swept and water-washed the floors, rid the place of a good deal of rubbish and opened wide door and windows to let the warm, salty air scour and scrub the foul atmosphere. After two days the place smelt sweeter, or we had become accustomed to the all-pervading sour smell of a farmyard muck heap.

That apart the cottage was acceptable; a true Hebridean two-roomed bothy, stoutly built in stone, gable-ended and roofed with corrugated iron; the walls were white-washed and the roof painted a gay red. And we, for a couple of weeks, were bothy men. The two rooms were spacious and dry; each had a fireplace, the walls close boarded, a concrete floor, and a tiny window which looked out on to a rocky foreshore. The view was always magnificent. In sunshine, here was paradise; when the wild, unruly winds howled in from the south-west the place was terrifyingly savage.

Two rooms were more than enough for our simple needs. One we used as a general store, the other as a combined living room and dormitory. The furniture was sparse: a substantial square table, the top covered with American cloth; two rough and ready wooden chairs made from driftwood, and a couple of wooden shelves hanging from one wall. The shelves carried our food supplies but not until they had been well scalded with hot water and lined with polythene sheeting. The table top, too, was scoured and scalded every day. Just simple safeguards to reduce the risk of contamination of foodstuffs. Flies were abundant; and there were rats.

On arrival day we did little about organizing camp. That could wait until tomorrow; we were far too anxious to look about us and make an early acquaintance with the flora and fauna. Goods were stacked impatiently in the store room, much-needed fresh air hurriedly introduced into the interior, camp beds erected, sleeping bags flung upon them, water fetched; and then we fled to the country.

In the late evening the Boy went fishing, came back all smiles with a couple of lye, one a 3-pounder. Lovely to eat, tasting like cod; big, fleshy fish with a three-pronged backbone as tough as steel and few other bones.

We turned in about midnight. The day had been superb—temperatures over seventy—and Roderick Cunningham had told us that this fine spell would last several days. During the night there was heavy rain.

Next day, which was a Sunday, we lazed away in idleness; that is, we didn't make a grand coastal tour or otherwise do the islands in one day. Instead, we cleansed the cot, organized our goods in the storeroom so that we could immediately find anything we wanted and not have to unpack cases and cartons in a wild frenzy, and in a leisurely way explored the local neighbourhood. The rain had ceased by breakfast, a meal we took about nine, and the day was another one of brilliant sunshine; consequently Roderick Cunningham's stock as a weather prophet rose considerably. In the night there was again heavy rain.

Throughout the twelve days of our residence the weather followed the usual island pattern, meaning that it was quite unpredictable and as likely to rain as not at any minute. We were alternately roasted by the sun, chilled by the cool winds, soaked by torrential rains. We basked beneath warm blue skies, cowered under cold grey ones. Maximum shade temperatures soared to 72°F on several occasions by day and sank below 40°F at night. The record low was 34°F on the night of June 21st–22nd. At 8.30 p.m. on the 21st the thermometer registered 44°F; twelve hours later the reading was 34°F, and there it stayed until noon when the sun came round to the west side and dispelled the chill. All times, B.S.T.

In fact, the weather conditions were not as dreary as that brief summary of temperature presents them. As usual, we began the holiday in warm, sunny weather. The 800-mile run from Kent to Shiants was near perfection, so, too, were the early days on the island. Then followed dull and windy days when we were glad to don woollies and thankful for our storm coats; back would come the sunshine, followed by another deterioration. Inevitably, the last day was a grey one and we rarely saw the sun again until we neared the Border on our way home.

Rain at night may seem propitious; when activity is confined to daylight, it is. Nevertheless, islands, where countless seabirds, come to breed, are well worth exploring after dark; certain birds, petrels and shearwaters, are nocturnal, and Shiant cliffs and beaches offer these species suitable breeding haunts. None has ever been reported breeding, but few ornithologists have made a prolonged stay on these bits of igneous rock. We never got round to late explorations, mostly the rain forbade. On a wet night bird activity is curtailed, and the Shiant cliffs are no place to be abroad when the night is wet and windswept. Our daytime excursions were always strenuous; long before nightfall, which wasn't far short of midnight, we were ready to stretch weary limbs and tired bodies on our beds.

We slept snugly in bags on camp beds and pampered ourselves with the added luxury of blankets, brought along in case the night temperatures dropped as low as they did. Our original plan was to sleep on the cottage floor, which we expected to be earthen, not the concrete that it was, but brought one camp bed, all we had, for emergency use should one of us feel unwell and crave for a sick bed to lie upon. At the last moment a friend offered us a second bed. We accepted, hesitantly—the interior of the car and the boot were already packed to capacity. As always, however, there was just room for one more item.

Many years ago, about 1925, Compton Mackenzie bought the Shiants and had the cottage either built or reconditioned for his own use. About twelve years later the islands came into the possession of Nigel Nicolson and he still owns them. For him they were a retreat from war-racked Europe during periods of leave in those dark years 1939–45. The grazing rights are let to a sheep farmer from Scalpay who keeps the cottage in repair and uses it during his visits to put down and take up his sheep, and at the lambing and shearing times. Nigel Nicolson very kindly granted us permission to land and stay on the Shiants and to use the cottage for shelter. Always I try to discover the owner of an island and ask permission to live upon it temporarily. I have never had a refusal and the owners have been most helpful, suggesting likely sources of boats, which will take me. This last leg of the journey can be very difficult to arrange.

In property advertisements the cottage would be described as

a roomy, country cottage ideally situated in protected surroundings on a low coastal flat, standing in own grounds and surrounded by a spacious paddock; (on a clear day) glorious
uninterrupted views of Harris, Skye and Scottish Mainland;
good fishing and safe bathing from private beach. Property
includes 200 acres of rock and moor. Seclusion guaranteed.

Company's water?

Well, no sir, not yet, but a spring, crystal clear and of great
purity is but 100 yards distant, cunningly hidden in bright green
grass and a profusion of wild flowers and herbs.

Gas and electricity perhaps?

Well, no sir, not as yet, but driftwood is reckoned to be fairly
plentiful upon the beaches, and peat may be cut, both free for
the carrying.

Sanitation?

Well, er, not exactly sir. We could sell you a spade. That would
be extra.

Refuse collection and disposal?

Ooohh! Chuck it in the sea and on the beach, everyone else
does.

Too true! Human beings are careless with their rubbish. The
beaches of an Tighe were cluttered with litter, some thrown
down by visitors, some washed up by the sea, brought from
goodness knows where. Women are as blameworthy as men;
empty cosmetic packs and other commodities peculiar to females
were common eyesores along the shore. There is no need for
this, disposal is so easy. Burial is one effective method. We learned
about another when on Handa. Emptied food tins can have the
bottom removed with a tin opener; the tin is then hammered
flat and all the tops, sides and bottoms made up into a small neat
parcel by putting them into a flattened cardboard carton, tied
with string. On the homeward journey the parcel is dropped into
deep sea. Please join my campaign. Keep desert island beaches
clean!

Amongst the general rubbish sometimes something unusual is
found. One of the strangest finds I made was on an Tighe, on
our next to last day. I had gone up to the cliffs at Campar on the
east side of the island in the hope of making tape recordings of
the kittiwakes that breed there, a hope that was blown sky high

by a gale-force wind. By the cliff edge, on the grass, lay the arm of a child's small dolly. I still have it, a souvenir from Shiant.

Behind the cottage is the ruin of an older cottage, and on the north side two more. Only the footings and small heaps of stones remain. Nowhere on Garbh and an Tighe are there any gaunt ruins, so common on other deserted Hebridean islands. Information about the Shiants is scant, although they have been visited and mentioned by Martin Martin, Macculloch, Muir and Harvie-Brown. The 6-inch ordnance survey map is no great help; it shows the outlines, high and low-water marks, and names a few headlands, beaches and small stacks, including the outliers. On an Tighe are shown the footpath, cottage sites, three wells—two of which are now fouled—and, in the south hinterland, the Pile. Near this Pile is the site of a church, marked by a few stones but otherwise indistinguishable and ignored by the map. Nor are the house sites of Garbh and Mhuire indicated, nor Mhuire's church dedicated to the Virgin Mary. The former inhabitants have long since gone, the last family leaving at the beginning of this century. Not really a family, merely a very old man and his daughter. Roderick Cunningham told us about them, how the woman would row 8 miles or so to visit her boy friend on Harris, or maybe Lewis, and when her father died she was alone for more than a week, unable to get off the islands. She must have been young then because at the time of our visit to Shiant she was still alive, living in Tarbert, Harris, and she had left the islands more than sixty years earlier.

A visitor's book was kept in the cottage, for Eileanan Seunta attracts a small number of visitors every year. Most of them are ashore only a few hours, some a day or two. From this book, which was disappointingly uninteresting and uninformative, we discovered that the more distinguished latter-day visitors included George Waterston and Tom Weir, although their visits were brief hours. To Tom Weir's bird list we could add only one. Once a school party had camped about the cottage precincts and made some enthusiastic claims, notably that the puffin population was 2 million. In passage were three young American girls who left their mark in the book and their litter on the shore. One wrote, "Happiness is moments." If her experience of Shiant matched mine she had lots of happiness. The most amazing entry was by two intrepid canoeists, who had boated over from Skye—the trip takes the Hebridean Ferry two hours—stayed a few days and were continuing to Stornoway, every bit of 40 miles. Naturally such intrepid travellers spurned the soft living offered by the cottage; they camped on the greensward 'twixt house and shore. Such a journey seemed impossible, but it was true enough. By coincidence I met one of the two a year later.

Except for sleeping and cooking we spent little time in the cottage. We fed extremely well; mostly from necessity out of tins, supplementing the canned goodies by freshly caught fish. In addition to such ordinary fodder as eggs, cheese, potatoes, porridge, bully beef, tomatoes and peas, our larder included delicacies such as lambs' tongues, ham, venison, chicken and salmon, soups in variety, peaches, pears, apricots and grapefruit, and a rich 3-pound Dundee cake, homemade. The big disappointment was the venison—one of the rankest chunks of mutton I have ever cheweed.

Our fuel was driftwood collected from an Tighe's shore, and calor gas in canisters burned in two small picnic stoves. Handa stew figured on the menu, but in small quantities as we hadn't a large pot for cooking. Bread we made, as required, from flour, butter, salt and baking powder, either as rolls or cakes, and cooked in a covered frying pan; time, about twenty minutes. The recipe is a simple one, and the making takes but a few minutes. Into a bowl—a saucepan will do—put, say, eight tablespoonfuls of plain flour; mix in a teaspoonful of baking powder; add butter, a piece about one-inch square, a pinch of salt, and mix well. The best way to do this is with your hands, until you have a good, greasy dough. You need cool hands otherwise the butter melts, and then you are in a mess. Having arrived at a proper mix add a little water and mix further. Within moments you have a sticky, gooey mess on your hands. This is not the moment for despair; persevere with the mixing. Sooner or later the ooze disappears and you are left with a nice doughy cake. Break this up and roll into small balls, any size you like, or flatten into a cake. Pop the balls or cake into a dry frying pan. Cover with an enamel dish or pan lid and cook for 20 to 30 minutes over a gentle heat. Every five minutes or so turn the balls or cake otherwise they will stick to the pan and burn. If you haven't a frying pan use a pot. Eat the same day, the bread soon goes stale. The butter isn't necessary, you may prefer not to use it. For milk you can use powdered milk mixed with water. We used real milk. On our journey northward we had spent one night at Dalmally, about 25 miles from Oban, and there we discovered something new to us: long-keeping milk, sold in cartons of one pint, and having a keeping time of ten weeks. We took half-a-dozen pints with us. You may not like the taste; we do. It is not that of fresh milk but not unlike that of tinned, skimmed milk.

What you would have liked, and more than a taste, was our wine. We took half a gallon, a home brew made from oak leaves, and matured three or four years. Making oak wine is one of the rites of spring. There are various ways of making it: using the withered leaves of autumn, using the young green leaves of June; or my way, which is to use the early leaf buds. In early May cut the young shoots, about 6 inches long, until a gallon measure is lightly filled. Place the shoots in an earthenware jar and pour on a gallon of boiling water and leave, covered, for twenty-four hours. Strain, next day, and discard the shoots; add 2 pounds of brown sugar, juice from three oranges and one lemon; put into a pot, bring to the boil and simmer for an hour. You now have the basic juice for your wine. From now on continue in the usual way of brewing homemade wines by adding yeast, allowing fermentation to take place, adding more sugar to taste, straining after a month, bottling and leaving to mature. We took our wine over in a polythene bottle. Polythene is generally supposed to have no effect on the contents. Don't you believe that. Water certainly is unaffected, but our wine took on a nasty taste; not permanent—after decanting and leaving for several hours the taste returned to normal. There is a faint sweet smell with polythene; the taste and smell are similar.

In addition to these essentials for living we had brought with us five cameras and auxiliaries, fishing gear, a small pocket wireless set, portable tape recorder and, of course, 'desert island discs'. These latter were on tape, a collection of music varying from 'pop'—for the Boy—through Strauss waltzes, classical jazz, Chopin nocturnes, Burl Ives ballads, Scottish reels and Shetland folk songs to selected movements from Beethoven and Dvorak symphonies. In all, three hours' musical entertainment. Although we played them every day, often twice, for twenty days—fourteen on the island, plus travelling time—we never grew weary of them. Indeed we still have them and often play them in the house in preference to the wireless or the gramophone. Our pocket wireless brought in Luxembourg loud and strong, but precious little else. It was useful as a check on our watches—heaven knows why, our days were timeless. There was never a tomorrow, nor a yesterday. And yet we couldn't forget our watches. Civilization has shackled us to the hands of a great ruling clock. Work for eight hours. Take an hour for lunch. Catch the 7.59 every morning five days a week; one minute before the train departs the barrier will be closed. Last train home 23.32. No more drinks after 23.00. London to Edinburgh by train in six hours; in one by plane. Save five. Tuppence for ten seconds 'phone time. At the third stroke it will be 10.52 and 20 seconds precisely.

Precisely.

Island life is unsophisticated, the day neatly sandwiched between daybreak and the going down of the sun; all that matters in between

> Eat when you are hungry,
> Drink when you are dry,
> Sleep when you are sleepy,
> Don't stop breathing or you'll die.

The author of that ancient piece of doggerel must have been a Hebridean; no doubt about it.

We compromised; a sensible course to follow. We rose, usually reluctantly, some time after we were fully awake. Our watches indicated this to be between 08.00 and 08.30 hours. If we were not out on the islands by 10.30 shocked conscience pointed accusingly at Time's face and urged us on. Hunger determined meal times; physical exhaustion suggested bed, always long past the ending of the day.

These late hours, bed after midnight, upset the nightly routine of one of the island's permanent inhabitants, the local rat. Animals are creatures of habit, their lives too are regulated by a clock. Ratty lived under a pile of stones, all that was left of a collapsed bothy, behind our cottage. We rarely saw him by day. At dusk he would emerge, slink past the side of the cottage, scuttle round the front and disappear. Perhaps he went down to the rocky foreshore. What he lived on we never knew. Very little edible matter was washed up; there were shellfish, winkles, limpets and barnacles. Living must have been hard; we never left any scraps inside or outside the cottage, all foodstuff in the cottage was under cover.

There may have been more than one rat but we are sure that there were not more than two pairs, and we never saw more than one at any time. After we had bedded down for the night, often before when we were quiet, we would hear feet pattering across the ceiling. Whether this was Ratty, or another resident in the roof, we never found out.

Our reconstruction of Ratty's night life was that he visited the cottage every night in search of food, experience having taught him that when visitors arrived food was to be had for the taking, easier pickings than on the foreshore. Our residence did not put a stop to his visits, only changed his timetable. The easiest way into the cot was through the entrance door, so we kept it shut, especially at dusk. There was another way, and Ratty knew about this—down the chimney. Both rooms had a fireplace and chimney and a door, so both doors were kept shut to prevent room-to-room access. In the living-room fireplace we lit a fire every evening, to ward off chills and for cooking such things as potatoes in their jackets in the wood embers, boiling water and heating soups whilst other goodies were being prepared over the picnic stoves. The fire was kept going until bedtime and then left to burn out. Not until the embers were dead did Ratty make his entrance, long past midnight. We saw him come one night. I'll say this for Ratty, he was a gentleman. He never trod soot all over the place, rarely left his droppings and did not touch our food, probably because he couldn't get through the rat proofing. Not once did he scuttle over me or my bed. I began to feel sorry for him—just a poor lonely rat seeking company—especially one morning at about dawn, when I awoke and saw him sitting on the window sill, looking out, quietly and forlornly. The Boy thought otherwise. His bed lay alongside the wall on which the shelves were hung. On these shelves were our food supplies. Everything under cover and rat-proof. Ratty regularly got on to the shelves and used the Boy's bed as a launching pad. Incensed at continued failure to steal our food and perhaps feeling in his bones that we were leaving, Ratty launched an all-out offensive on our last night. Outside, the weather was wild and windy, rain beat on the iron roof, and the wind howled like a pack of Dervishes. The shore was no place for a rat on such a night. About two o'clock the Boy awakened, felt something on his feet, guessed it was Ratty, and kicked. Ratty sailed through the air and landed plop, on the Boy's chest. For a while there was pandemonium. We never saw Ratty again. I often wondered what his reaction was next evening.

We knew about Shiant's rats before we came. Any mention of the Shiant Isles in literature has a reference to the rats. Martin Martin appears to be an exception. The islands, state various authors, are overrun with rats, and the boulder field on the west shore of Garbh is mentioned as their main stamping ground. Here, too, is a large puffinry, said to be equal at one time to the puffinries on St. Kilda, but considerably diminished because of the invasion of rats. This boulder field is approximately half a mile long and 200 feet wide. The depth is considerable. Amongst these boulders nest a great many puffins, who prefer the rocks to burrows. We found only a small number of puffin burrows but repeatedly saw puffins loaded with fish disappearing into the boulder jungle. Counting them on the boulder field was difficult, at the burrows it was so easy. Our rough estimation was 20,000 pairs. Living with them were razorbills, friendly little birds, and when we sat quietly they would stand only 3 or 4 feet away, shuffling and nodding to us. We never saw a rat nor any bird corpses or skeletons; not even a puffin turned inside out, the trade mark of the greater black-backed gull. None of these great birds frequented the puffinry. Perhaps the rats too are diminished; the winter living must be hard when the sea fowl have gone. Do the rats abandon Shiant, swim across to Harris and Lewis led by a piping wren to a shore of plenty? Maybe we were rat blind, and great armies still live in the talus, spending the summer killing and storing puffins for winter days ahead, dragging their carcases to the cottage for monstrous orgies. And Ratty is king, the cottage his palace.

The only other mammals we saw on Shiant were sheep; shepherdless, as is common on lonely Hebridean islands. These were not scrawny Soays but a cultivated breed put down to grow wool

(*Above*) Newly hatched shag, bald, blind and black. (*Below*) The shag, a bird of coastal waters, nesting on clifftops and wide ledges

Treshnish. (*Above*) The southern half of Lunga Island is flat-topped.
In the distance are the two Dutchmans. (*Below*) The west coast of Lunga
and the Harp Rock, centre of huge colonies of guillemots, puffins,
kittiwakes and shags

and become good mutton. Some never stayed their term; whitened bones, a heap of wool, a fly-ridden carcase marked their
untimely end; hideous feasts for the black backs and herring
gulls, hoodies and ravens and, maybe, the rats.

A small tribe, four ewes and two lambs, had attached themselves to the cottage, breakfasting and dining regularly on the
pastures; sometimes stopping for lunch. They regarded us with
that peculiar mixture of tolerance and suspicion which sheep
bestow on human beings; kept a safe distance of several yards
between them and us when we approached slowly, and threw
caution to the winds when we appeared suddenly and without
warning, kicking up their heels and skimble skambling in every
direction but ours; then made an equally sudden halt and turned
on to us a bleak and empty stare, silent and accusing.

Ours was a pastoral world; a cottage on the shore, sheep in
the meadows, larks in the sky and the murmur of the sea on the
rocks beyond making soft music. A place to laze away summer's
days. Occasionally a basking shark cruised by, a shadowy thing
in pellucid waters, only a dorsal fin and half a tail clearly visible.
Once in a while a seal lifted its head, gazed sorrowfully around
us and sank again. Perhaps it was looking for a lost mate, lured
away by some cunning human, a seal wife deprived of her skin;
perhaps there is more to these ancient tales than we shall ever
know.

When the tide went down the foreshore was left as a maze of
rock pools, some deep; we used them as our baths and for washing
smalls. At first I went through morning ablutions gasping and
shivering in an icy sea. Urging the Boy not to neglect personal
hygiene I was told, pityingly, that *he* used the rock pools. By
evening the water temperature had been raised to comfort level
by solar heat.

Our nearest neighbours were herring gulls, a small colony of
about twenty pairs, nesting along the west shore of an Tighe and
about 300 yards from the cot. Without fail our arrivals on their
territory were heralded by a great commotion amongst the adults,
dive-bombing attacks to discourage our trespass and continuous
mewing and muttered warning cries to tell the chicks to hide.
The first cry went up when we reached a spot about 200 yards
away, gradually increasing in intensity as we drew nearer, and

8

reached a frenzied climax when we halted by a nest. Increasing family responsibilities made these critical days.

On this side the island tumbles down to a rocky shore and low cliffs, rarely more than 20 feet high; below them are rocky platforms, some completely covered at full tide. The rocks above high water are clothed with yellow lichen which, in the sun, shines like burnished gold.

Halfway along the shore is a miniature ravine, the stepped and grassy hillside dropping steeply to a narrow corridor from which rises an uneven wall of rock. The battlements are thronged with gulls. About here is a mixed colony of herring and greater black-backed gulls, up to a hundred pairs of herring gulls, and closely adjacent a dozen pairs of black backs. Further on, at the southern tip of an Tighe, called Mianish, is another greater black-backed colony, thirty or forty pairs strong, nesting amongst the rocky outcrops and cushions of thrift, a typical habitat for this fierce gull. Fierce but not always courageous. In my trespasses into greater black-back territory I have rarely been dive-bombed. Mostly the greater black-back flies off to a nearby elevated vantage point, muttering angrily and sulkily. In like circumstances the herring gull does not hesitate but dives steeply, a feathered fury intent on destroying the intruding enemy.

Opposite the point of Mianish and separated by 30 or 40 yards of sea is Sgeir Mianish, a low mound of rock, a preening station and fishing lookout for shags and gulls.

The highest point of an Tighe, which is at the Mianish end, reaches to about 400 feet, not a sharp peak but a rounded top sloping gently away east and west. The east cliffs are straight-sided, columnar and ruggedly buttressed, falling a couple of hundred feet on to rocks and sandy, shingly beaches to which access is very difficult. Only at one place did we find a possible way down, so uninviting that we did not attempt it. These cliffs are almost empty of birds. At no place are there ledges for guillemots and razorbills, and the cliff tops, where we expected to find puffin burrows, were deserted. In a tiny cove near Mianish was a small colony of kittiwakes, probably fifty pairs. These and the gulls on the west shore were the major portion of an Tighe's bird population. Mixed with the gulls, in isolated situations, were a few dozen shags, many with eggs, some with young in the brown woolly down stage, and one eider duck. Oystercatchers screamed at us whenever we visited the west shore, and we often saw them probing amongst the rocks and reefs; occasionally we disturbed a snipe feeding in a boggy patch, a lark or two rose suddenly, pipits darted and wheatears flew from rock to rock, sometimes silently, often scolding continuously.

The eider's nest was cunningly hidden amongst rocks—it needed to be as this was herring gull territory. The gull enjoys making a meal of other bird's eggs and the chicks. This horrid trait ought to deter eiders, and other specie, from nesting within gull territory, but doesn't. Most eiders are trusting birds and, unless taken unawares, do not leave the nest until the last possible moment. Those that leave early are probably young and in-experienced birds; I have noticed that these nervous eiders look young—that is, the appearance of the mandibles and lores are fresh and clean, not hoary and sere. This particular bird was unconcerned about our presence. We came upon her unexpec-tedly, approaching from the front in her full view. She sat unmoving and as she was so serene we retired a little way to erect cameras and tripods, then photographed her at close quarters. Her nesting site was marked so that we could watch progress but on our next visit we reached the nest unexpectedly and suddenly, and from behind her; off she flew to reveal one hatched duckling and two others partially hatched. The day was hot and sunny, the chicks unlikely to come to harm for want of warmth. The gulls were the danger.

Eiders do not readily return to the nest; they approach slowly and deliberately in stages. We took up a position amongst the rocks above the nest and watched through binoculars. If gulls came along we could deal with them. The eider was some time absent, almost an hour, and by then the hatching was complete. She had gone down to the sea and spent a long time swimming, then flew to a rocky perch and preened her breast feathers. After more than half an hour's absence she slowly walked over the rocks in the direction of her nest for a distance of some 20 to 30 yards, halted for several minutes, doing nothing in particular, made another brief walk and paused again before flying to a ledge from where she could see her nest. Unexpectedly she did not fly direct to the nest but stood waiting patiently. Something was disturbing her because she paced up and down slowly, looking round her very carefully, and finally flew to a vantage point near us. She must have seen us for almost immediately she flew off and alighted very near her earlier perch; and there she stood for many minutes, apparently undecided about her next move. As stealthily as possible we crawled away and left her to her vigil. When we returned much later she was at the nest brooding the chicks. Next day she was gone, but an eider with three little chicks was swimming contentedly inshore.

Several pairs of eiders were attached to the Shiants. Some days we would see them off the west coast, on others in the Bay on the east side, several families together. Eiders are amicable birds, swim happily together the day long, and any duck will readily look after another's ducklings and adopt a lost one. Only once have I seen an eider duck attack one of its fellows, and that was at the Shiants. One evening a party of four ducks and a dozen ducklings swam along by the landing beach, searching the seaweeds, at times almost swamped by rough seas. One duck, swimming alone, a short distance from the main party, was repeatedly attacked by another; eventually she swam off on her own.

The duck alone incubates the eggs and looks after the young. The drake is a resplendent fellow in white and black, with a touch of pink on his breast, black on the crown of the head and with patches of pale green on the nape. At nesting time he flies off to a remote place to moult and is not seen at the breeding station. His plumage, in eclipse, is dingy and he does not regain his former glory until early in the following year.

The numbers of eiders to be found about small Hebridean islands are few; up to six families swimming together is the maximum I have seen, and they have mostly nested apart, singly, not in a colony as they will do in Iceland. On these small islands their choice of a nesting site is near to the shore, sheltered amongst rocks, but I have found an occasional nest several hundred feet above the sea. When possible the nest will be concealed by vegetation. On one of the Summer Isles there was a nest deep in the heather, only the duck's head visible and that only seen when I knew exactly where to look. Incubating eiders sit stone-still not even blinking an eyelid when you stand a few feet away. The eggs hatch more or less at the same time—four, five or six usually—

and next day the duck takes her brood to sea and there they stay together, coming ashore at evening time to roost on a beach or amongst the rocks for the night.

The *Handbook* states that the duck takes the young down to the water as soon as they are dry; those I have found waited at least several hours, up to twenty-four. As these nests have been in or near gull territory, not the safest of places, maybe the duck has waited until dusk, or darkness, before going down to the water. Most Hebridean eider ducklings are shortlived. There are many hazards: rough seas waterlog them or sweep them away from the main party; along comes a gull, sometimes a hoodie, and snaps them up. Sometimes a greater black-backed gull will swoop and seize a duckling from the sea. The duck puts up a spirited defence, all too often unsuccessful. Day by day the brood diminishes; a sad sight to see the number dwindle. Often none survive.

The Boy rescued one waterlogged duckling at about midnight, washed ashore and entangled in seaweed, a bedraggled bird with little life in it. We took it into the cottage, dried and warmed it by a driftwood fire. Rapidly it recovered, became once more a lively ball of fluff. We detained it overnight; to keep it snug and warm I wrapped it in a woollen sock and put the sock and chick into a deep side pocket of my storm coat, fastening the flap buttons to prevent the bird's escape. For a long time the wretched creature kept us awake, chirruping loudly, non stop.

Just before dawn I awoke and saw, near the window in the pale light of Hebridean summer night, a small dark shape moving. Ratty! I thought, and was about to heave a boot at him; and didn't. Rats don't chirrup. I never expected the duckling to escape from the imprisoning pocket. To prevent a possible tragedy of the rat-eats-duckling order, I scooped up the bird, dumped it into a cardboard carton, with a sock or two for warmth and comfort, covered the top securely and left the bird to chirrup the rest of the night away; which it did. Next day, when a couple of eiders and their families sailed by, we released our lodger in the sea. Within seconds he was re-united with his own clan.

In the afternoon of that day the Boy found another water-logged duckling, lost, in need of care and attention. We were eider farming. Here on the Shiants, bringing a new lease of life and much needed commerce to this remote island group. Soon we would have thousands. There would be pens all over the cottage meadow, and stretching down to Mianish; great batteries of deep litter; free range too. Ships would put in for eggs. We would have a stall, slices of roast eider in home-made bread, washed down with flagons of oak-shoot wine. And just as we were growing rich along would come a revenue cutter, put an excise man ashore, questioning about undeclared earnings, evading taxation; shocking scandal; life-long imprisonment; or banishment, to North Rona; there we could start again, farming Leach's fork-tailed petrel.

Our second patient was released the same day; one night alone with an eider duckling is enough. Number two we released on the rocks in front of the cottage, and recaptured him, or her, two or three times. We demanded payment for treatment. Nothing excessive, just a little acting in return for kindness shown. We filmed him running across the rocks, making his glad way back to the sea. He swam rapidly through an expanse of calm water, scampered over a small raft of seaweed and into rougher water. One of the adult eiders saw him coming and swam towards him. When they met, the adult bent her head forward and gently touched the little one on the crown of his head—a pretty scene. And then the pair swam away to join the rest of the party.

Beyond the flat pasture at the back of the cottage rises a vertical wall of rock, unclimbable because it lacks toe and finger holds. Access to the humpy, steep-sided hill which lies above the wall is from a point near the wall. Water from the well floods the ground about, creating a tiny bog in which grows water mint and forget-me-not in profusion, orchis, buttercup, meadow sweet and a few bits of willow withy. The earthy hillside shelters more flowers, especially in the hollows, and from the summit opens up bird's-eye views of Mhuire and the east coast of Garbh. The sides of an Tighe's east cliffs are invisible from the tops; they fall sheer, tremendous straight-sided columns of basalt, on to a fore-shore of rock and shingly beaches, just about accessible at low tide by scrambling madly over and round the tumble of rocks at the southern end of the storm beach. Seaboots are recommended, or go barefoot if you can stand the tingling cold of the water.

The landing beach is about 100 yards long, and some 15 yards wide at normal high water. Exceptionally high water, caused by strong winds and storm conditions, narrows the width to a few feet; in winter the beach is probably flooded. During our stay the winds were mostly westerly, ruffling the waters on the west side of the Islands and leaving the bay wondrously calm. A very strong wind would ripple the surface of the water in the bay and send spindrift skating across the surface as it blew the tops off the wavelets that buffeted the rocks close inshore by an Tighe's cliffs. When the wind reached gale force the west-side waters were lifted right across the landing beach into the bay.

The herring gulls were quick to take advantage of this. On most days there were a few fishing down by the beach, sailing low or swimming in a leisurely way until fish chanced to come their way. One evening we descended from Gharbh's high tops in the teeth of a gale shrieking in from the south-west, the western waters lashed into a whirlpool of white foam that would be the envy of any whiter-than-white clothes-washing detergent. The bay was unbelievably calm, and swimming in it, close into the storm beach, were the opportunist herring gulls, forty or fifty of them, making quick forward sallies as the wind caught up the sea and threw it across the storm beach into the bay. Trapped in this tangle of wind and water were fish. Some dropped on to the beach, some into the bay, to be quickly snatched by a gull; some reached neither, the gulls dextrously caught them in mid air. The herring gull is not a great fisherman and will eat a variety of foods; it is never slow to take advantage of easy pickings.

Gharb Eilean is a hillier and rougher island than an Tighe, topping 500 feet, and a naturalist's paradise. The hillsides are a mass of flowers, commonplace and unexpected. In the swampy places grow cotton grass, yellow flag and marsh marigolds; secluded suntraps on the cliff shelter primroses, violets and blue-bells; heather and bracken clothe the hill tops; pink thrift grows amongst the rocks and on the cliff sides, sharing the thin soil with campion. The meadowy places not yet overrun with bracken are carpeted with daisies and buttercups, birdsfoot trefoil and tormentil; orchis are as common as the daisies. Scotch thistles, not yet in bloom, were thick with great, green flower buds.

Unexpectedly we found canes of a wild rose, not yet blooming, and looking very much like the English dog rose which is reputed to be rare in Scotland—maybe it was sweet briar. And, as if this was not enough, larks filled the air with song; pipits and wheatears carrying food in their beaks tantalized us, urged us to seek their nests; starlings added a homely touch; and the cliffs were thronging with seafowl.

Our favourite excursion was to the east side of Gharbh, reached at low tide by scrambling over the seaweed-covered rocks that skirt the great cliffs. Our objective was a huge boulder field; boulders as small as a football, as large as a motor-car, bigger than four or five 'buses, collapsed into a mountain of stone, a gritty jungle covered with yellow lichen, shining in the sun like polished copper and gold. Down in the scree live thousands of sea fowl. The shags nest in the cavernous openings on the outer surface of the boulder jungle; associating with them are a few guillemots. The main bulk of the inhabitants are razorbills and puffins. Shags were counted only in hundreds; puffins, which nested deep down, were there in thousands, and for each puffin there were three razorbills. All used the outer boulders for perches to preen, and laze, and fidget.

Encounters with shags were sometimes alarming. As we squirmed in and out of the labyrinth we would come unexpectedly upon a shag, sitting on eggs or alongside its chicks. The surprised bird launched itself from the nest in wild dismay, an awe-inspiring sight. On its feet the shag is an awkward creature. Leaving the nest hurriedly its feet are pounding and wings are outspread, flapping powerfully to assist locomotion. If the bird is a cock it is barking hoarsely in protest at unwarranted trespass into its seclusion; cock and hen flounder out of the rocky shelter wildly, rather like a novice pilot making a bad and bumpy landing on a rough, grass runway, or a power boat on full bore leaping through a choppy sea.

The chicks squeak shrilly and if big enough will leave the huge, seaweed nest and back away into the darker corners of their dens. Unlike many birds of the sea and coastal waters shags are born bald and blind; black skinned, ugly little creatures; a young pterodactyl probably bore a striking resemblance. Only its parents could love a baby shag. They grow quickly, in a few weeks are in a browny down, and fly when about two months old.

Occasionally we caught puffins and razorbills as they scrambled out of their rocky hideaways and into our hands. They were never perturbed about their capture, they did not react violently. We kept a close watch on their beaks, for these are strong and can administer a sharp nip. The puffins made no attempt to bite, they lay quietly and contentedly in our hands until tossed into the air, then flew serenely off to sea for a flip around the bay. Two razorbills took a revengeful nip on a fleshy part of a finger. The Boy was bitten the previous year, at St. Kilda; a Shiant razorbill showed me the power of his bill. The bites were not severe—a sharp pinch which broke the skin, not painful although a long time healing, several weeks elapsed before the mark finally disappeared.

On the edge of the boulder jungle and beneath a small jumble of rocks we found an adult puffin with a newly hatched chick at a crevice opening. The chick was not long out of the egg, still wet, sticky and trembling. This was our first sight of a puffin chick. The evening was windy and cold, no sun warms this side of Gharbh in the evening, so we marked the spot and left parent and chick to snuggle together for warmth. When we called on a later day we could see the chick in the tiny cavern, inaccessible and out of camera range.

By the time we left Gharbh the tide was in and had cut off the approach route, so we made our way back over the top of the island, climbing 300 or 400 feet up one of the few passes in the wall of rock that makes Gharbh all but impregnable on the eastern and northern shores. From the cliff tops there are magnificent all-round views of the distant hills on Harris and Lewis in the one direction and Skye and the mainland in another; nearer at hand, Mhuire, to be seen distantly, or seemingly only a stone's throw away when viewed through glasses. As we crossed from east to west over Gharbh's flowered crown we could see the whole expanse of an Tighe, from the sheer cliffs climbing off the storm beach, to the Mianish crags a mile away. In the foreground nestled the white-washed cottage on its green plateau, looking as small as a shoe box. Away beyond it, near gull territory, the ridges of long abandoned lazy beds stood out bold and clear in the evening sunshine; to the east of them was a boggy bit, a forest of yellow flags and a haunt of snipe—we often disturbed them feeding but never found a nest. Then the island climbed gently to its highest point, resembling, from our high viewpoint, an old-fashioned sofa, a giant's chaise-longue.

The descent from the top of Gharbh to the beach is steep, through gullies. There is a choice of two paths. In places they are wide, and then narrow to a few feet. The surface varies. Here it is grassy and slippery; there, stony and rough. Bumpy, grass-grown slopes are interrupted by giant steps. A long, rough and rocky staircase, once gated, opens on to a tumble of rocks by the beach. When a gale blew, standing on the exposed heights could be exhilarating. Sometimes we didn't walk down the bumpy slopes but slid on our sit-upons, a quick, exciting descent, very bottom bruising too—occasionally blood was drawn.

Once we were caught at the top in a sudden storm. The prelude was a quickening wind; in moments a gale was blowing. The darkening sky went inky black. The opening raindrops became a torrent. The furious wind lifted our raincoats over our heads. In a couple of minutes we were soaking wet. No good seeking the shelter of rocks, I told the Boy, this lot's here for hours. The cottage is our haven; let's get there. The descent was a nightmare. The wind bashed us, and the rain lashed. When we reached the beach we were so wet we might have crawled out of the sea. As we crunched across the stones the winds abated, the rain stopped, dark clouds stormed out to sea and left the sun in an azure sky. Half an hour of sheer hell, and now a heaven in all the glory of silver and gold, sapphire and emerald.

On serene evenings we would amble down the hillside back to the cottage and, arriving, prepare an evening meal over a drift-wood fire: warming soup, lambs' tongues, ham and pickled onions, potatoes baked in their jackets, peaches and cream, biscuits and cheese, a hunk of Dundee cake, coffee and a goodly measure of oak wine to settle a contented stomach—all taken to the soothing strains of our desert island discs.

Whilst the meal was cooking we'd enjoy a refreshing tub in a rock pool, and, after the platters were emptied and washed, laze away a late evening watching the eiders come inshore to feed. As the sun went down behind Harris the Boy would fish for to-morrow's breakfast.

He began by using limpets, collected from the rocks on our domestic shore, and discovered that the fish rarely took the bait. He was float fishing using a river rod and a light line, and found that the fish, lye, were attracted by the bobbing float; by substituting a spoon, or a feather, armoured with a hook, he caught all we needed. Most catches weighed about a pound, the bigger fish were getting away, all too often with spoon, line and hook. The line was only 3 pound, never intended for serious sea fishing.

On most evenings we were entertained by a magnificent aerial display over Gharbh's eastern headlands, thousands of birds circling the towering cliffs. The performance would begin at about six, gradually building up and reaching a peak of maximum activity at about ten, seemingly every shag, razorbill, puffin and guillemot taking part. The sky was dark with birds, the scree packed with them—temporarily parked, we assumed, whilst they regained strength—a continued coming and going. The massed formations in the sky flew in every direction: east to west, west to east, to north, to south; in straight lines, in circles; some high, some low; climbing, swooping; masses of puffins, hordes of razorbills, swarms of guillemots; formations of shag flying boats, singletons too; down for a breather, up for another dizzy spell on the merry-go-round. Not a gull joined them; nothing but a great circus of auks and shags. At eleven, when dusk spread across the islands, the birds began to come into the cliffs for the night. Some late roisterers whirred on until past midnight. From the cottage we had a grandstand seat at this amazing circus. Only at St. Kilda have I seen anything to compare, where the great puffinries on Dùn and at Carn Mor on Hirta provide a like performance. The Shiant exhibition seems the more spectacular, probably because it is on a smaller stage and the density more concentrated.

The ascent of the great headland of Garbh is strenuous going. Resident sheep find it no great obstacle; I wonder what those fat and pampered creatures of the lush, rolling downland would make of such wind-skimmed islands. All the two breeds have in common is the vacant stare. Hebridean islands are not places "where sheep may safely graze". One false step and sheep is carrion. Big surprise is that this hillside, in the days when the Shiants were inhabited, did not deter the islanders' cattle from seeking places where the grass grows greener, for we have it in Martin Martin's own words that "the Cows pass and repass by it safely, tho' one would think it unsafe for a Man to climb".

This is not Garbh's highest top. That lies half a mile onward above the solid northern cliffs, reaching to 528 feet. Nor is this half mile an easy saunter over flower-strewn meadows. Garbh undulates, is a real switchback island, very much so on the western side. The eastern end of the north-facing cliffs drop down to a small, green oasis, complete with lochan and burn— favourite resort of sheep—then reaches out to a rocky promon- tory with a natural arch, Toll à Roimh, and a deep cleft which almost severs the point, Bidean à Roimh, from the main bulk.

Beyond the low cliffs that edge this wee pastureland of Airid- hean a' Bhaigh is a reef, submerged at high water, and a small stack Sgeirean a' Bhaigh, a miniature lagoon, and, at low water, happy hunting ground of oystercatcher and gull. About here there is a cleft in the columnar rock face up which, like the cows, we passed and repassed—very steep, very breathtaking, very good for leg and thigh muscles. Not recommended for those with high blood pressure. In the wet, very slippery; when ascending one wrong step and the bottom is quickly reached; there you recover, unwind yourself, and try again.

We were clambering skywards late one afternoon, with a storm pending, when the sudden screams of annoyed fowl caused us to pause in wonder. For a full minute there was the noble sight of a golden eagle sailing majestically over the shore, harried by a screaming oystercatcher and an arrogant greater black-backed gull. There was no time to fix the glasses on the big brown bird; not until it had gone did we realize that this wasn't a buzzard as at first anticipated; it was larger than the black back. We never saw another. This could be Shiant visiting time for golden eagles. The date was 20th June 1967. In the diary at the cottage was an entry by Tom Weir dated 21st June 1963, stating briefly: golden eagle seen.

Our 21st June, celebrating the first official day of summer, was more in keeping with approaching autumn, Strong south- westerly winds blew fierce and cold, drove many heavy rain squalls over the islands, chopped up the sea, and put the thermo- meter down in the forties. Mid-summer day, 24th June, was even worse; a day of drizzly rain, mist blotting out Harris and Lewis; hardly a breath of wind. The disenchanted islands. A day for indoor sports. I made bread and brought our diary up to date. The Boy repaired torn trousers and dried out driftwood before a roaring fire. We played our desert island discs twice over, and dined well; kept the home fire burning brightly, banishing Ratty to the cold, cold shore, or his shelter beneath rusting corrugated iron.

The next day the weather relented, it was the Sabbath. How wonderful to get abroad again, in shirt sleeves, in sunshine. We spent the day worshipping it.

Garbh, as an Tighe, was high, sheer cliff on the east side and dips low on the western seaboard, a crinkly coast line extending a mile from the storm beach to Stocanish. There the cliffs mount high and turn a corner, the rock face broken and ledged which the guillemots have exploited to the full; this is their main breeding station at the Shiants, and not easily observed from the cliff top.

We spent an afternoon and evening following the shore, exploring every little bay and cove. The greater part is rough and rocky. The exception, half way along and close by the sheep fank, is a lovely, green bank, sloping gently to the sea and owned by gulls, herring predominating, a few greater black backs on the flanks, and two pairs of lesser black backs. An ideal spot to lounge a while and sip coffee, discounting the protesting gulls who were affronted by our intrusion and said so. For a reason that puzzled us for some time a pair of lesser black backs dive-bombed us throughout our stay. The gullery presented the usual scene: scattered nests, downy chicks, leggy chicks, some hiding, some sunning; adults sailing, perching, growling, snoozing. We sat by a half-buried boulder for twenty minutes enjoying the view and the coffee. As we rose and turned to go I saw a bright glint in a crevice of the rock where we had lounged. Here was the reason for the vicious attacks by the lesser black backs, for behind this rock, in an earthy hideaway, were three gull chicks. The glint I had seen was the sun on a chick's eye.

Before we had got this far we had narrowly escaped an oiling by a fulmar whilst climbing round a high, grass-grown, earthy bank, 50 feet above a barren, rocky floor. There were few secure footholds and only chancy hand-holds. Unexpectedly we found ourselves at a point demanding the utmost care in negotiating and face to face with an incubating fulmar. The bird rose on its tarsi, drooped both wings, drew back its head and partially opened its beak, sure signs that a fulmar is putting its artillery into action. We halted, hoped that our precarious foothold wouldn't give way—it didn't; hoped that the fulmar would delay action—it did, but remained at firing station. Very slowly, very gingerly, we crept by. A jet of hot fulmar oil is a nasty thing to have hit you; so is a rocky floor coming up fast.

Dismay! When we rounded the corner and were out of fulmar range and view we discovered we could not go on, the earthy cliff was unsafe. Back we came, trembling unhappily, once again to face a fulmar in ambush. We made it unscathed. I wanted the Boy to brave the fulmar's wrath whilst I filmed the episode. He wanted me to do so whilst he filmed. I began to understand his reluctance.

Leaving the unfriendly gulls we walked to the sheep fank and put to flight a mixed party of adult and juvenile starlings, searched the walls of the enclosure for nests and found none. On the slopes above the fank we found our only evidence that oystercatchers breed on the islands, a solitary well-grown juvenile, rooted to the ground in the shelter of a rock. When the Boy picked it up the bird remained motionless, unconcerned. Then he sat down, holding the bird in his lap imprisoned in his hands. Mum and Dad were circling the hillside, screeching their heads off in shocked alarm, and Junior remained quite still, unflinching. Next, the Boy set the young bird on the ground alongside his leg; it sat, as if on the nest, unmoving, for several seconds. And then, in a flash, it was on its feet, spun round and ran helter-skelter to a new hiding place.

On the moorland above the far end of the west shore, Stocanish, we found a colony of breeding lesser black-backed gulls, about fifty pairs, with chicks running free, scattered and hiding in the rough grass and reedy beds. On the homeward trek, as we neared the storm beach, a hen blackbird and a couple of juveniles flew up from bracken.

No other visitors had recorded blackbirds in the diary, and this was the only new species we could add. All others had been seen and recorded time and time again.

On the tenth day of our stay the wind went round to the south-west, worked up to gale force and put the sea into a great commotion. Next day a storm broke and for forty-eight hours we were lashed by driving rain and whipped by a salty wind. Tumultuous seas broke over the rock piles and frothed their way beyond the foreshore. White foam surged around the feet of the cliffs. Clouds came down from the hilltops and met the sea mist at the beach. We watched a shag flying into the gale from the bay. As it rounded the headland to cross the storm beach the full force of the wind struck. The shag hung in the air, its large wings beating strongly and rhythmically, powerfully holding its own against the gale but not able to beat it, and finally had to give way and was swept back into the shelter of the cliffs.

When the storm abated we had to leave. We were expecting the *Isle of Skye* to arrive at noon. At 9.30 a knock summoned us to the cottage door. Roderick Cunningham was waiting, and we were packing with leisure. We sent him away on an hour's cruise, and, miraculously, finished packing on time. As we carried our many packages down to the beach *Isle of Skye* came back into the bay, stood off and put down her boat to retrieve us. Overhead puffins and razorbills whirled, shags sailed and the gulls screamed. We went aboard, the boat was lifted in, and under grey skies we headed for Tarbert. The sun greeted our arrival.

Five hours later we were on Skye, enjoying the luxury of a steaming hot bath. What a wonderful thing is hot water. When you've been without it for two weeks, civilization has nothing better to offer.

SIX

Alone with the Wild Sea Fowl

·⊹·

"The greatest difficulty is getting there from Mull. Another way is from Iona", wrote Lady Jean Rankin in reply to my letter asking permission to live on her Treshnish Islands for a short while.

"If you have no means of transport for getting between the Islands, Lunga is the best to stay on, being the largest and having fresh water. Fladda also has fresh water. There are no buildings left on any of the islands. There are many specie of seabirds; unfortunately the puffins seem to be less in numbers each year."

Lunga it was to be, and a tent, approached at about nine on a calm, June evening. The loveliest of islands with all the beauties of the Hebrides concentrated on its 170 acres. A green jewel of an island in a peacock sea. A paradise, a Garden of Eden, mine for two weeks. No Eve, just me, and the wild sea fowl.

Once more on a golden afternoon in mid-June I pointed the bonnet of the car to the North, rolled out of Kent, dived under Thames and emerged on the Essex flats. This year there was a difference. I was alone, the Boy could not come. He was turned 17, no longer a schoolboy, now stepped up to an engineering apprentice, a mucky, manual worker, his weekend's brightened by a small pile of shiny brass. Sad that he couldn't come; I would have to do all the humping myself. Familiar land marks slipped by: the horrid intersection at Brentwood; the long detour round Cambridge; shoulder to shoulder through the narrow ways of Huntingdon. Freedom at last, zipping up the Great North Road. Tomorrow I would be in the high hills of Scotland; the day after, sleeping on Tertiary basalt.

One of Alistair Gibson's boats put me on this very basalt at about nine in the evening, seventy-five minutes after we had left Fionnphort at the southern tip of Mull. The sea was dead calm, the landing so easy. The little boat slid up to a rocky promontory, and there, deck high, was a ledge. I stepped out on to it, took my goods piece by piece as they were handed to me, and stacked them into a hollow at the top of a stairway. No dicing with icy death, waiting tensely for the right moment, then a mad leap ashore, on to an all-too-narrow ledge.

A landing place, a preening station, a fishing look-out, that was the purpose of this rocky arm. No spot to pass the night; even the sea birds shunned it after dusk. Solid rock too. I wanted a mattress of soft, grass grown earth between me and the basalt before I bedded down. At the shoulder of the arm, only 60 yards away, there was such a camp site. Back home 60 yards is nothing —why the end of the garden is farther distant. On an island Nature contrives to place every conceivable obstacle in the way. At two-thirds the distance a stairway led down to a wide platform; on the other side more steps climbed an earthy bank, at the top of which was a grassy landing just wide enough to take a small tent. Good enough for a night's camp; tomorrow I would reconnoitre. First things first and priority number one was the erection of a tent.

I had brought a small ridge tent which, the makers claimed, would sleep two. Who was I to contradict such an authoritative statement especially as in the catalogue was a diagram, to scale, showing two bods lying side by side. Quite so. But what do you do with your possessions? Take a second tent? That is what I should have done.

There was no shelter for my twenty pieces of goods. This night I stowed in the tent items that must be kept dry, clothes, tape recorder, cameras and their auxiliaries. Other packages remained in the open; some were protected by black polythene sheeting and dumped alongside the tent, the rest were left in the shelter of the rocks on the promontory. Tomorrow I would be organized. For the moment I was weary and forty hours behind with sleep. No time to go exploring, the velvety purple of a Hebridean summer night was wrapped around Treshnish.

Mid-afternoon is the ideal time to go ashore. An hour or so

9

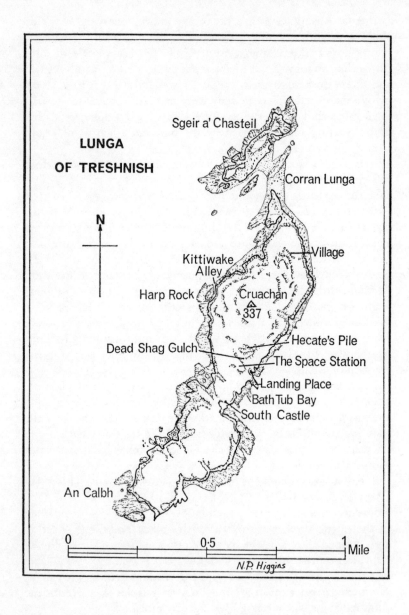

Sgeir a' Chasteil

LUNGA
OF TRESHNISH

Corran Lunga

N

Village

Kittiwake
Alley

Harp Rock

Cruachan
△
337

Dead Shag Gulch

Hecate's Pile

The Space Station

Landing Place

Bath Tub Bay

South Castle

An Calbh

0   0·5   1 Mile

N.P. Higgins

spent portering; another hour in trifling organization of camp, just the bare necessities, somewhere to sleep, something to eat and drink. A little desultory unpacking of certain essentials such as binoculars, 6-inch maps and camp spade; then a leisurely reconnaissance of shore and meadow, flora and fauna, location of the well, a sweeping search of the uplands through the binoculars. The evening to be spent in gentle relaxation, lolling in the sun, pampering yourself because you have had a long and tiring journey, and in your head road, tyre and engine noise is still echoing, beating out the miles—all 700. And then to slip deeply into soft, refreshing sleep.

Ugh! Morning so soon. I could have slept the day away. The sun was above Mull's distant hills, flooding into my tiny cove, warming my tent; and the sky was hazy, promising a day of shimmering heat. Before the sun and the temperature climbed too high I fetched the rest of the goods and took stock of the camp site. One immediately obvious objection was that the plot sloped in two directions, and in no way could a tent be pitched so that it was in a comfortable horizontal plane. There was a place, above the bank which sheltered me and 200 yards away, that offered possibilities; a wide and flat expanse of short grassland. I never moved there. The great deterrent was the bank, high and steep, up which all my belongings would have to be carried. Additionally, the outlook from the tent would have been bleak; the immediate surroundings were featureless, a hummocky moorland with stone outcrops, and no view of the sea.

The ledge was an idyllic spot and there I stayed. Above and around me, in the steep banks, were puffin burrows; below, on wide recesses in a weather-worn cliff, 20 to 30 feet above the sea, the noisome dens of two families of shags; down in a rocky hollow, on the platform that separated me from the landing place, a herring gull had her nest; and high on an earthy, inaccessible cliff a solitary fulmar incubated her egg. All about me were rocks yellowed with lichens and fringed with shags; niches in the banks sheltered primroses and violets, the rims dressed with sea campion; at my door spread mats of silverweed and birdsfoot trefoil. On these I could laze, look across the sea to Mull, Ulva and Gometra, reach for the glasses and search for a speck called Staffa. Early in the morning the sun wakened me, warmed me as I sat at my front door breakfasting; regrettably it never stayed to bid me goodnight. Towards nine in the evening it was dipping slowly towards the western seas, throwing a deep, cold shadow across my estate.

My first day was one of small misfortunes. When unpacking, the thermometer was found to be broken. As I lazed on Costa Lunga dressed only in my skin, the binoculars slipped off my knee, slid a yard and were pulled up sharp by a small rock; the jar was sufficient to knock one prism out of alignment; until I returned home I now had only a monocular. The Hebridean sun, so often tempered by the wind, scorched me badly, my shoulders, my arms, my chest, my legs. So many times we have packed anti-sunburn lotion and never needed it. This year I forgot to bring some; I burned for days. More serious was my failure to find a water supply; the sunburn had given me a thirst that would have brought untold riches to a publican.

At Fionnphort Alistair Gibson had told me that the islands were very dry and no great amount of rain had fallen during the past six weeks. He would mark on my 6-inch map the location of a well (but forgot), insisted that I took drinking water with me and put aboard 2½ gallons, for which I was to be very thankful. At the time I wasn't unduly concerned. That the Hebrides and the west coast of Scotland could be rainfree for several weeks seemed to be beyond the brink of possibility. A casual look round the island, however, showed that Alistair Gibson was right, Lunga was dry. What had once been water pools—and were now depressions of dried and deep and widely cracked mud—bore silent testimony; burns that had gushed sparkling, peaty, brown water were parched gutters.

The 6-inch map did not mark a well. Near the ruined village was the obvious place to search. I combed the place square yard by square yard and never a well did I find. Not until I was led to it seven days later did I know the location. There was an old static tank full of clear water amongst the village ruins, and this I used, after boiling, for washing dishes and me. My drinking water was severely rationed to a pint a day. Fortunately I had brought eight pints of long-keeping milk and that was a tremendous help; nevertheless I had to be careful in the choice of food, eating nothing salty or otherwise thirst-provoking.

Cooking was done with difficulty. I had brought the two small gas-fired picnic stoves, but unless these are used in a shelter they are not satisfactory; the wind cools the saucepan faster than the flame can heat it and so the contents are neither cooked nor raised to any degree of warmth. After I had found a shallow and small wooden box, washed up on the rocks, and just large enough for the two stoves to stand in, and built a protecting drystone wall around them, I did manage to boil water and heat milk and soups, but only if the wind was gentle. A stiff breeze found its way through chinks in the stone wall. There were no beaches nearby and rarely a piece of driftwood to be seen; and so I was denied the alternative of a camp fire. When the winds were strong, in desperation I wriggled into the tent with a stove, swept out of the way impeding kelter, and boiled water there. Sometimes the situation bordered on the impossible; then, in despair, I gave up cooking and subsisted on dried and cold foods, longed for a wee stone cot where I could spread myself, and drowned my sorrows in oak-shoot wine. In my sourest moments I decided never again would I use a tent, a contraption which is useful only for people who have no possessions other than the clothes they stand in, which must be worn day and night through-out the duration of the expedition, and food limited to strips of pemmican carried wrapped around and hung about their person. When morale was high I made bread scones, piquant soups and tasty stews and eyed the puffins tentatively. The St. Kildans reckoned them good to eat, would toss a dried puffin on the fire and leave it to toast; a great pity there was no driftwood. When I was in these high spirits even the baby shags moved nervously about their smelly dens.

The shags were my nearest neighbours. On rising in the morn-ing I would step across the few yards to the cliff edge, bid them good morning with a friendly wave and descend the staircase to my favourite pool for a bath. They never approved of me as a resident; even the youngsters gaped rudely, writhed their scraggy necks and protested shrilly. One morning there was only one chick in the nest. The adults never came back, and after an hour or two of loneliness, during which time he or she paced the narrow confines of the nesting ledge agitatedly, the bird decided to join the rest of the family at sea. I watched it go. For a time it shuffled about the ledge, peering over and about, then retired into a dark corner to hide, all the time squeaking shrilly. Finally it toppled off the edge more, I am sure, by accident than intention, floundered over a nearby rock or two, scrabbled along a narrow ledge, fell headlong down a miniature precipice, fetched up short on a rock table and slid down uncontrollably into a shallow inlet to experience its first taste of sea water. With that baptism the bird recovered its equipoise and swam further along the inlet to the face of the rock, where, after a long interval, it was found by the adults. Why it never broke its neck in the nest-leaving remained a mystery. Shags must be made of india rubber. No wonder they are considered inedible. Brave—or desperately hungry—is the fisherman who will cook and eat a shag.

My other neighbours never protested. The puffins regarded me with a mixture of curiosity and old-world courtesy. Charming old ladies and gentlemen in evening dress, who would gather on the bank, look down on me, cock their heads on one side, look at each other, occasionally rub bills, sit and stand in leisurely array and seemingly ask each other what's *h. sapiens* cooking this evening, and what's that strange plaything of his that makes weird noises? Chopin soothed them, for they sat unmoving. Beethoven, Dvorak and Strauss would have them cocking their heads, and once, when the Foula Reel started up, several of them got on to their feet. Only once was there an exodus, when a throaty bass boomed out a folk song; perhaps the puffins thought it to be a greater black back.

The fulmar up on the earthy cliff was completely indifferent. She had an egg to incubate and there was nothing more important in the world. Occasionally her mate would fly in and they would sit and cackle just as though he had recounted some doubtful joke picked up from the rude shags. When either of them left, he or she flew away, never zoomed about the local cliffs and bay in a typical fulmar way. I had a feeling that they didn't belong, were only here for a season; colonials, perhaps, from some far-off sea, sent over, with a grant from their government, to the Hebrides to carry out oceanic research.

The herring gull too, was uppity, did not seem to have a mate, only a nest with an egg, which was an encumbrance, and three favourite perches. One was on the rock spur where I landed, another on the cliff near the fulmar, and the third on a ledge immediately in front of my tent and only 20 feet away. All three overlooked the nest, and the third perch was only occupied when I was away. As I returned I would see the gull fly off. Twice, whilst hidden in the tent during a ritual with the picnic stove, I watched the bird take up a position on this point. Five days had passed before I discovered the nest. I was skipping down the staircase and across the rock terrace to my bath tub when the gull flew up from some secluded nook, and there I found a nest with one egg from which a small piece of shell had flaked off. During my stay the egg never hatched, nor were any more laid. The bird did sit on its egg occasionally; once or twice I surprised it in the very act of incubation, and then it would fly off to one of the favourite perches and remain there in what I supposed was meditation. This was a lonely bird, it could have been an outcast. Never did it make any attempt to mix with other gulls, the nearest of which were 200 yards away. Whether it was a hen or cock I never knew; I called the bird *Dumbelle*, because it never uttered a cry, and I had a feeling that it was a hen.

Lunga is a small island of 172 acres, approximately one and a quarter miles from southern to northern extremities, and 600 yards across the widest place. The northern half is humped, "a pile of eroded flows", and rises to 337 feet. The southern half is flat and from the high top is seen in its entirety, a great and green terrace lying 80 feet above the sea. The edges are ragged and deeply indented where the cliffs have eroded. In two places the terrace is interrupted by wide ravines, which slope to the shore, ending as stone beaches. During my brief occupancy the higher levels of the beaches were never washed by the sea. When winter's gales shriek in from the Atlantic these beaches probably receive a tremendous drubbing; high up on the green floor of the ravines were large and heavy pieces of driftwood, planks, tree trunks, pit props, thrown there, no doubt, by tempestous seas running before storm wind.

The cliffs on all sides of this southern part of Lunga are rugged and colourful, not negotiable without ropes, not stratified into balconies, not earthy topped at their summit and hillocky, no place for any auk. Between east and west cliffs is a savannah, mixed grassland, marsh and bog, the grass surprisingly green in this time of drought and populated by wheatear, pipit, greater black-backed and herring gulls. At the foot of the cliffs are rock pavements along which I could walk at low tide, gaining access to them through the ravines. Down here, sheltered from winds, I lazed away hours watching the busy oystercatchers searching the reefs, and the shags preening, spreading their wings and drying feathers in the wind; there was always a fringe of them on the rocky spurs that break the waves and beat the sea into a froth of white water. Gulls, too, came to fish; occasionally an eider and her family passed this way.

The northern part of the island, mountain country, is very nearly all hump, correctly known as the Cruachan, and springs almost off the very edge of the east and west coast cliffs. The approach from east and south is by gentle slopes; the northern and western sides are steep, falling in a series of terraces, each supporting the one above on vertical walls of rock, broken here and there so that man and sheep can scrabble from one to the other, often with difficulty. This contouring of Cruachan is not very noticeable when clambering about it; to be appreciated fully it must be seen from one of the other islands in the Treshnish group.

From the top of Cruachan the whole group is in view, from the two Dutchmans in the south-west to Cairn na Burgh More and associated Beg in the north-east. Lunga is roughly midway between, the overall distance being 4 to 5 miles. Nothing but Atlantic ocean separates Lunga from the Dutchmans; in the 2 miles of seaway between Lunga and Burgh More and Burgh Beg are fifty, plus or minus a couple, islands, islets and skerries, a series of giant stepping stones. Excepting the larger Dutchmen and Lunga, all are flat topped.

On one of the northern terraces is the ruined village, deserted now 100 years, the cottages stone skeletons, a typical island sight. I was surprised to find the village this much, 80 feet, above the shore. The villagers must have had a wearisome time fetching and carrying to and from the beach; the way was long and steep. There must have been a good reason for building in this place, although it was not obvious. The close proximity of fresh water, perhaps, was the deciding factor.

Lunga's north-western corner does not end in an abrupt vertical wall of rock. Here is a beach of sorts, playground of gulls and oystercatchers. Where the cliffs turn the northern corner from the east the rock wall has broken down and crumbled, detritus sloping to meet a stony beach; and a flat nose reaches out, shingly, stony, grass-grown, edged in one place by a low platform, the remnants of a cliff. Beyond is Corran Lunga, which can be crossed on foot at low tide to reach the islet Sgeir a' Chasteil, 100 yards away. The visit must be short, an hour including the crossing, as the incoming tide quickly floods the narrow passage. There is little to see: two flat-topped eminences, one at each end, another about the middle; 600 yards from end to end; lounging station for sea birds; beachcombing for waders; somewhere to fly to, somewhere to fly from; next nearest land, westwards, Coll and Tiree.

All the wild flowers of the Hebrides seem to grow on Lunga, Nature's great garden and horticultural centre. I was bewildered by the colours, the varieties, the quantities. What Garbh Eilean of Eileanan Seunta had to offer, Lunga could offer three times over. All the common flowers had a showing, the locally common, the uncommon, the scarce plants which grow only in limited areas. On the sheltered sides of the ravines clumps of primroses and violets bloomed; on the open terraces were huge beds of birdsfoot trefoil; buttercups and daisies encroached by common right. In the miniature corries on Cruachan the wind tinkled clumps of bluebells; at its foot on the western side was a field of orchis, growing so close that it was impossible to walk without crushing them under foot. The cliffs were hung with sea campion, sea pinks and little alpine-like flowers that I did not recognize. Wild rose was in bud, honeysuckle about to flower, brambles carried green buds, scotch thistles too. Cotton grass danced in the bogs; yellow flags brightened a dull corner by the north-western beach; tormentil, silverweed and dandelion were not uncommon. There were flowers and plants I did not recognize and had no time to discover their identity. One such find was growing on a beach near the mouth of a ravine—a prostrate plant of grey fleshy leaves, a great mat of it, with a bluish flower, just this one clump. Later I learned that it grows only on one other of the Treshnish Isles—which, I do not know. Eventually I was able to have it identified from a coloured photographic transparency I took. Oyster plant is its name—so called because the leaves taste of oysters. The plant is scarce and decreasing, growing on coastal shingle in Scotland and very rare elsewhere in the north. Quite a find; I wished I could have appreciated it at the time and not weeks later.

Hebridean flowers know no division of season, spring and summer are as one, beginning late, ending early. Having enjoyed the beauties of the early spring and summer flowers in Kent there is double pleasure in seeing them all again, together, not weeks apart but gathered in one great flowering mass on less than half a square mile.

I discovered ferns too, in a silent cavern, hard by the sea. A dark, mysterious, spooky place, little known and unsuspected. Something more than a cavern, a tunnel through the rocks into which light never penetrates, and where no sound is heard, not even the wild wind, nor the sea, not even a murmuration of the faintest kind.

This place of mystery lies in the southern terrace on the western side. I had seen it as a dark mark on the green when looking down from Cruachan, and came to it curiously, full of wonder. At close quarters it was revealed as a circular pit, 20 to 25 feet deep, 50 to 60 across, the walls straight sided except at one place, and here was a steep slope down which I scrambled to the floor. Once below the rim there was an unearthly silence. The vegetation was rank; there was no beauty, nothing to encourage exploration, only a curiosity that had to be satisfied. On the seaward side of the pit I found an opening, a cave of some sort. The roof was 6 or more feet high, the floor loose and dry. I shuffled my feet and brought to the surface shells, 2 or 3 inches across, like giant cockle shells. There were lots of them, all buried, some 3 inches below the surface. No indication of what could have brought them there. No signs of rats or rabbits, no droppings to prove they lived in this place. Nor did I find bats. Just the place for a hoodie to nest in, perhaps a raven or rock dove. No sign of nests, no whitewash, no signs that this was a dining hall of some beast or bird of prey. Nothing but shells and silence. I went in deeper. The path widened, rose and fell;

the last glimmer of daylight faded away. Then the path climbed
and the roof came lower, I had to go down on all fours to pass
through a narrow archway. The silence was oppressive. Ahead
was a glimmer of light, the path declined, the roof was raised
and I came into a small cavern of multi-coloured rocks, and near
the entrance, where the light was stronger, grew a few sorts of
ferns in niches on the cavern walls. Now the floor climbed again
and was blocked by fallen rocks, easily scrambled over to return
into full daylight. Beyond were flat rocks, seaweed-covered reefs,
the sea shuffling over them, an oystercatcher searching them;
above, blue skies, fleecy clouds, and a brassy sun; blowing in, a
boisterous west wind; a world full of noise again. My torch
wasn't powerful enough to uncover any beauty the tunnel might
have. I took a couple of photographs in colour with flashlight,
and these reveal the colour of the rocks and the rough hewing.
Whether the tunnel is a work of nature or man-made, and for
what purpose, I've never discovered. Only once have I found a
reference to it, and that by accident. I took off a shelf in a library
a book describing Staffa and turned to the index idly wondering
if there were any references to the Treshnish Isles. There was
and, to my surprise, mention was made of this tunnel; no details,
just that the ravine nearby is called the Dorlinn, something I
didn't know until then.

Despite the brave show of the many-hued flowers, and the
verdant pastures, there was no doubt about the lack of water.
The marshy, boggy flats were parched, the moorland pools
empty, their floors dried and criss crossed with wide cracks, hard
as stone. Probably sufficient rain fell for the vegetable kingdom's
needs, but never enough to refill the pools. Certainly there was
rain during my stay: heavy, squally showers lasting half an
hour; occasionally in the night and early morning there were
periods of non-stop, steady rain, once for a few hours. Nevertheless, Lunga's water pools remained dried and cracked. Long
ago I concluded that when the folk of Western Scotland and the
islands contended that there has been no rain for weeks, they
refer to continuous rains. Intermittent rain, even half a day of it,
is dismissed as nothing. I don't think the people grumble about
the lack of rain; they are simply amazed at its absence. Some
appear to be appalled at its presence. When we came down from

Handa we stopped overnight at Ullapool. The day had been a poor one: dull, little sunshine and rain in the evening; rain beating down monotonously next morning. The lady who had bed and breakfasted us lamented: "Oh! the rain, and this wass Chune; such a wet month it had been; this wass not Chune, oh no!" Perhaps the lament was to comfort us, an apology for so inhospitable a climate. We know what she meant. The Hebridean Chune can be so lovely; this one was.

There was rain; not in great quantities. My diary of the weather reads "Day three, rain in the morning. Days four, six and seven, some light rain. Day eight, many prolonged heavy showers (I remember those). Day nine, occasional heavy showers in the morning. Day ten, rain squalls. Day eleven, some rain." Could an island-goer in the Highlands expect anything better? Summed up, my weather was occasional rain showers, sometimes heavy. Never did they seriously inconvenience me. There was far more dryness than wetness. The Scottish summer of 1968 will be forever remembered for the sunshine.

I saw only three mammals on Lunga—mouse, rabbit and sheep —either distantly or a glimpse so fleet that it might have been a shadow, a trick of the light.

Of sheep, that common element of the islands, there were only two; the grazing tenant had lifted the rest of his flock. The truants had eluded capture, roamed Lunga wild and defiant, living the life of bandits, alert and ever watchful, seeing without being seen. Our paths never crossed. On the day of my departure, as we sailed inshore, I swept Cruachan with the glasses, and there, amongst the crags, lurked two sheep.

Total count of rabbits was one. Widely scattered patches of droppings and closely nibbled pastures were evidences that the population was, at least, a few pairs. I never saw rabbits in the evenings, which is a favourite time to step abroad, graze and frolic in warm, greeny hollows. They could have seen me lumbering towards their playgrounds, and fled. More than likely my journey back to camp did not take me through their territories. Usually my path skirted the southern and western foot of Cruachan, across rough and tussocky moorland—nothing juicy and succulent, no larder for a rabbit.

The mice were a surprise. Between the tent and the flysheet I had stored a piece of cheese in a polythene basin and discovered one morning that a corner had been nibbled by what looked like mouse teeth. In the basin were a few mouse-like droppings. Field mice, I decided, and regretted that I hadn't any catch-a-live traps. The cheese excepted, no other food was ever touched; up until this discovery no mice had been seen or their presence suspected. The cheese was left for the invisible visitors. Later, I sensed more than saw a dark flash of movement in the grass by the tent. The mice never disturbed me at night. I slept unconscious of any micey games being played upon me; no tail ever tickled me back to consciousness. They, or it, came into the tent. One was disturbed near the head of my bed. A momentary glimpse was all I had, a dark form with a whisking tail bolted and vanished instantaneously.

Later I was to learn that these creatures were in all probability house mice, descendants of the mice that had lived in the village cottages more than 100 years ago. In 1937 Dr. Fraser Darling was living on Lunga near the village; mice soon attached themselves to the new and temporary inhabitants and made so free with the expedition's stores that "trapping was essential and we caught more than 75 individuals". And still they came. Darling states that these were house mice living as field mice. His mice were resident in their old field habitat within 50 yards of the collapsed cottages. My visitors were at least a straight half a mile from their place of origin. Their journey from the northern half of the island to the mid point where I camped would be nearer to a mile, the path is a circuitous one and for mice an adventurous migration.

Several species of seabirds colonize the cliffs of Lunga; guillemots, kittiwakes, shags, puffins and razorbills are all numerous. Fulmars, greater and lesser black-backed gulls and herring gulls are present in much smaller numbers. Fulmars nest here and there. No big colony, but small scattered communities on the east cliffs; there is room for many more.

Largest of the cliff dwellers are the shags, to be seen all round the island—a shaggy fringe, as it were, on the cliff edges, the skerries and the reefs. Puffins and razorbills nest on the east and

west sides and are absent in the north and south. Guillemots and
kittiwakes are concentrated in one particular area on the west
side. Here, too, is the biggest shag colony. This is birdlan, a
great metropolis of thousands of birds of the ocean and the
coastal waters, come together for a few months of the year. By
late summer the noisy, thronging cliffs are silent and empty.
Puffins, guillemots and razorbills, adult and juvenile, have gone
their separate ways, far into the ocean, to winter in green seas,
with only the wind for company.

The rhythm of life around my camp was leisurely. A few
dozen birds close at hand; a few hundred a step back. I had my
own pipers, a pair of oystercatchers who lived in the tiny bay
below my camp. I called it Bathtub Bay because when the tide
ebbs pools are left, shallow and deep, some several feet deep,
filled with seaweeds, tiny fishes, molluscs and shudderingly cold
water into which I plunged daily lest I outpower the smelly
shags; then a brisk towelling and a short skip around the rocks
so that the sun could air my birthday suit. Above and beyond
the bathtubs, and above the reach of the sea, is a terrace of black
rock, the rough face softened by sea pinks and shallow saucer-
like hollows into which the wind has shifted coarse sand.

Access is by a not-too-difficult path around the cliffside, a
nimble jump across a small gap—wide enough to add a tremor
of excitement—to land neatly, sometimes in a heap, on the
opposite rock. Then merely to step up on to the terrace and find,
unexpectedly, a scrape with three eggs, the property of my
pipers. Always step warily in these places, they hide precious
things. Protective coloration merged this nest into the surroundings so well that I stood within a yard of it for some
minutes before realizing it was there. Once the location was
known there was never any difficulty in spotting it again, not
even from a nearby earthy headland towering 80 or more feet
above the shore.

From the headland I hoped I might be able to watch the
oystercatchers' activities at their nest, and photograph them; I
had 400-mm and 800-mm telephoto lenses with me. Every
attempt was foiled by a greater black-backed gull which lived on
the tip of a much lower and opposite headland. The bird was an
alarmist spoil-sport. I would creep to my vantage point, some-

times squirming along on my tummy, and no matter how stealthily I moved just as soon as my head appeared at the cliff edge, "Owk, owk owk," cried the black back, away went the 'catchers;' and as long as I lay there Claptrap "owked" his-her head off.

One day I invaded his-her territory. He-she didn't stop to fight, like a bold herring gull, just flew off to a rock to mutter sullenly. The nest was now abandoned, a focal point only. Three chicks lay huddled together on the short grass, and separated at my approach. They were several days old, no longer pretty, fluffy creatures but at the ungainly stage, leggy, beaky and their bits of wings sprouting feathers. There was no escape route for them, they could only fall off the headland on to the rocks and into the sea below, so I turned back to save them from instant extinction.

On the lower slopes of this headland two herring gulls had nests, one with chicks, one with eggs. Quiet birds showing only uneasiness at my presence, no rumbustious bellicosity. Claptrap, Dumbelle and these two herring gulls were isolationists, living away from the main colonies.

The greater black backs have three main colonies of twenty to thirty pairs each, two occupying exclusively the eastern and southern parts of the island. The third and smallest colony share the north-western coastal flats with herring gulls and a small colony of lesser black backs. There is an exclusive herring gull colony in a cove near the Dorlinn, enclosed by high cliffs which serve the birds as a clubland and from where they can keep watch on their breeding territory below.

About twenty-five pairs nest here, a rubbish heap of a place made ugly by green and slimy pools, flies infesting decaying sea-weed and the remnants of rotting fish. This noisome pit faces the sea but is not directly exposed because pavements, rocks and reefs, low and tall, give some protection. Narrow and broad seaways wind about the rock and through them. On a windless day, water heaves and sighs as it gently and rhythmically swells and subsides; on a gusty day when a get-out-of-my-way wind blusters inshore there's slosh and wash as the fast ebbing water of a spent wave is hit by the next incoming buster. All's commotion, a great muddle of water, skirling, whirling, froth and

bubble. The surprise is that no water kelpie, no spume-flecked
water-horse bursts to the surface to drown unsuspecting and
innocent ornithologist. There probably is a toll taken of young
gulls.

As I watched, a gull of about ten days lost its balance on a
rock and toppled into calm water. The little bird swam well
enough, chose to leave the pool and make for a narrow channel
between two large rocks, where it could have stumbled ashore.
Had it not been for an untimely wave the bird might have done
so. Instead, a sudden rush of water picked up the hapless creature
and bashed it against the rocks. Fast following second and third
waves battered and swamped it again. Then, drawn by the suck
of the ebb the now helpless bird was thrown into even more
turbulent water, the sea shuttlecocking it from rock to rock. What
tough little creatures are young gulls; for nearly a quarter of an
hour the sea played with it—and the bird survived. Finally it was
tossed to a place I could reach. I grabbed it, a wet bedraggled
mass of feathers, dead beat. I dried it and brooded it and when it
opened its eyes again and showed signs of life, lay it on the grass
in the sun to warm and dry out. Ten minutes later it was almost
a gull chick again, stood up and shakily walked off to seek its
own hideaway.

Leaving the gullery I too came to grief, walking along the
beach, carelessly hop-skipping from boulder to rock to boulder,
stepped on a rounded stone, which tilted. I crash landed. Bruises,
minor cuts and a broken watch. For ten days I lived timelessly,
and how it irritated—more than the cuts. I had not brought a
wireless, and I found myself telling sun time. I knew by earlier
observation that when my camp was in shadow the time was
20.00 G.M.T., that sunset was 21.30. Time didn't matter and yet
in ignorance of time I was befooled into thinking that I was being
cheated of a birthright. I should have been glad to throw off the
shackles, to live by simple time, not by the hour. I am awake,
time I was abroad; darkness comes, time to sleep; I am hungry
and thirsty, time to eat and drink; that is a beautiful scene, that
bird is behaving strangely, time to take photographs. What
exposure? Oh! hell, we're back to split seconds.

Time to return to my camp and carry on describing my
neighbours. Around Bathtub Bay are three lots of crags; they

Treshnish. One of the several hundred kittiwakes nesting on the sheer
cliff face near the Harp Rock, Lunga

Hoodie crow's nest and chicks

Treshnish. (*Above*) The Space Station, colonized by puffins, razorbills and shags. In the centre is the author's camp. (*Left*) Leaving Lunga. Alistair Gibson, in his boat, takes the author's possessions from the landing place

stand like ancient castles and all are climbable without much danger to life and limb. One has extensive cavernous dungeons at the foot wherein lurk shags, razorbills and puffins.

South Castle is of modest size and stands on a bluff beyond Claptrap's territory. Care is needed when promenading round the front; a treacherous pathway by the cliff edge, no second chance to correct a false step—just a long and nasty drop into the sea. The path ends abruptly by an even nastier drop; so face about and return. Not a place to visit during and immediately after heavy rain. Nor at night. A few razorbills and shags are resident. Hoodies use the ramparts as a lookout. On the front face in a small cleft I found a wren's nest, still being used by the fledged young; four flew out within inches of my head.

The crags behind me, The Space Station, are more impressive and extensive, with supporting outbuildings. There are large gardens and terraces front and rear, open to the public, every day and all day. Could have been a royal castle once, the palace of a kingly eagle, long since murdered, his line brought to extinction by a bigoted gamekeeper, the noble heads suspended from the keeper's disgusting gibbet.

Imperious shags have taken over, the sentinels barking a 'who goes there' whenever I passed and repassed. In the caverns hide their queens and princesses, hatching piles of pale blue eggs into black reptilian monsters. A newly hatched shag has to be seen to be believed. Squadrons of lumbering shag space ships are constantly launched, flights of puffins and razorbills too. For ever scramble. Enemy sighted—me—approaching on photographic reconnaissance. Alarm over, bandit gone, return to base. Bring fish, Princess Shagreen's hatched another egg.

The third castle is aloof, a standing ruin on a broken hill, the battlements in fallen disorder. In daytime a mere collapsed cliff. In late evening, that sinister time when long shadows creep stealthily across the hills, the seafowl shun the place. Hecate's Pile. Occasionally a raven circles. At dusk hoodies flap slowly around the top.

> How now, you secret, black and midnight hags!
> What is't you do?

I would never have been surprised to see steamy vapours rise from a hidden cauldron. Do the hoodies "round about the cauldron go" weaving spells from a sticky vat simmering

> Egg of gull, and puffin chick,
> Beans of Heinz to make it thick,
> Wine of oak to give it kick?

I ought to have gone there on midsummer midnight and watched Birnam Wood move in, but I didn't know when midnight would arrive. My watch was broken.

Below the Space Station is a pleasant cliff walk which descends into a quiet, unfrequented hollow, strewn with fallen boulders. Under one pile was an isolated shag's nest, monarch of the glen; under another pile a dead shag, more smelly in death than life. To my delight, after an hour of patient watching, I found a wren's nest, cunningly hidden behind an enormous tuft of withered grass that hung from a cliffside and resembled a Santa Claus beard. I first saw the wren standing on a boulder, tail bobbing, grubs hanging from its bill, disturbed by my wanderings about its territory. I froze and watched. Stage by stage the bird flew to the cliff face and disappeared, reappearing seconds later minus the grubs. The wren was quite tame; as long as I was still, the bird ignored me. During the grub-hunting intervals I got to within 15 feet of the nest and could see clearly that the entrance was beneath Santa's beard. After the next visit I moved over and looked. Nothing! until I lifted the grass tuft and found the nest entrance a good 18 inches higher. Inside were four nestlings, wing feathers emerging. A very tame bird this wren. I placed a camera openly within 15 feet of the nest and photographed arrivals. The results were acceptable. A wren is a diminutive bird; at 15 feet, even with a three magnification telephoto lens, the image is small.

On the far side of Dead Shag Gulch a steep bank climbs to Hecate's Pile. Turn left for Cruachan's lower slopes, clad with stumpy bracken; bear right for the east cliffs crowned by undulating turfy banks, one the site of a small puffin colony, others gathering places for shag matins and soirées. Scattered over the cliff faces are nesting fulmars, mostly out of sight from the edge.

A bird I have long wanted to see in its nest is a puffin chick, an uncommon sight because it lives at the end of a long burrow.

There is a way, dig the burrow; not as you would dig out pota-toes, for this would ruin the burrow and be unkind to the puffin. The digging must be done with care. Always included in my pile of goods is a garden trowel which, with a camp spade, is suitable for excavations.

Up until my visit to Lunga my curiosity had not been satisfied. Earlier attempts were failures; either the burrow chosen was unoccupied, or I got lost in underground labyrinths. A burrow the Boy and I examined in the Shiants simply led us to another entrance. In time we found that three other entrances simply led to the other two and each other. Confused, we gave up. On this hill on Lunga I had watched a puffin enter a burrow; when I looked in the bird was just inside, and shuffled further along. An arm thrust in made the bird edge further inside. I decided to dig.

Before digging begins, flat stones must be collected; they are never near at hand. These are needed to re-roof the burrow. Digging must be done slowly and carefully. The turf layer is skimmed off with a spade and laid aside for later replacement. Now a vertical cut is made through the burrow roof, near the entrance—a narrow slot, not the full width of the burrow—and followed along the burrow length. Quite often stones have to be removed, leaving a gaping hole. Fallen earth is trowelled out through these holes to prevent blockage of the tunnel. The puffin meanwhile, if it is in the burrow, moves further along until it reaches the end; when it remains stationary this signifies that the burrow end is near, and digging stops, otherwise the bird might be hurt, or the egg damaged. The puffin has a powerful beak; gloves are a good protection. My puffin was reluctant to leave; as it happened there was a chick and the adult was probably protecting the little one. The parent was lifted out and stood on the bank, where it shook off crumbs of earth and went headlong to the sea, probably for a wash. The young puffin was only a day or two old, the egg tooth was still at the bill end. The chick is nothing like its parent, just a fluffy ball of black down, a straight beak and substantial legs and feet.

Puffin chicks spend the first six weeks of their life in the dark, and dislike the light. I put mine in a dark place, and examined the nest; nothing more than a scrape in the earth, a couple of feathers, a few bits of grass. The excavation was finished by cutting a circular hole in the earth over the nest scrape and closing the hole with an overlapping flat stone. The tunnel was now cleaned out, all loose earth and other debris carefully removed; and the flat stones placed along the tunnel top to cover the cut away portion. Next the turf was replaced and pressed into position. All I had to do was replace the chick in the scrape, close the opening with the stone and put back the turf. On future inspections all that was necessary was to lift turf and stone above the nest, peer in, reclose.

As a check against nest desertion, a couple of pieces of light stick were placed crossways at the burrow entrance, and when I returned that way hours later, they were down. Subsequent inspections in the following days showed that the young puffin was alive and well and continued to thrive.

I was on very good terms with the puffins about my camp; however, their burrows were not suitable for excavation because they were well down the bank not on the hill crest and could not be opened up from the top.

Puffins at their breeding station are tame. My local puffins were very tame. They quickly accepted me as part of the cliff colony—a sort of flightless, mateless, unconventional old bird living in a peculiar burrow it had somehow built, and who could moult at will in warm sunshine and grow thick, new feathers when the wind blew cold. As long as I moved slowly and quietly I was tolerated and could sit within 3 or 4 yards of a party and watch them. Soon I began to recognize some of them as individuals, small differences in size, plumage and behaviour characteristics marked them. Bighead had a noticeably larger head, Hoppity limped, the Caretaker had an aged look and was for ever picking up stones, pieces of rubbish and vegetation; Bellyflop either had short legs or middle-age spread, he hadn't much ground clearance.

Ashore, puffins spend many hours in idleness, standing or squatting on the cliffs enjoying the sunshine and sea breezes. Some actions are very quaint. The mincing promenade is one. The bird stands high, drawing up the body, stretching the neck upwards, and pulling the bill in to touch the lower part of the front of the neck; at the same time it takes a few forward steps lifting each foot high above the ground; to the human eye the

birds look most ridiculous. At other times, when the bird seems
to have some definite object in view, it will waddle forward,
head and neck lowered, with a right old seadog's roll. They play
a game of billing, not as doves and budgerigars will do but rather
like a couple of dogs muzzling each other. The game rarely lasts
more than thirty or forty seconds because other puffins join in;
this stops the two contestants who look at the other puffins,
haphazardly wag their heads, break off and wander away. I filmed
one or two sessions and have tried to analyse the game, not very
successfully. There aren't any rules. A couple of puffins casually
moving around meet and consent to play. One pushes its bill
towards the other, who cocks the head slightly, pulls back,
pushes forward, and their bills make contact; heads begin to
circle as they rub. There is no rough play, all is very gentle
sparring, and one bird rubs along and down to the lower man-
dible of the other, then inverts its head and proceeds to rub the
base of the opponent's bill; there is no conclusion because by
then other puffins have drawn near and are pushing in their bills,
as though to say, "Come on mate! have a rub with me." The
several puffins look at each other, jerk their heads, and the party
breaks up. Minutes later another bout starts. Billing doesn't seem
to be an appeasement ceremony, simply *bonhomie*. In these idle
gatherings puffins act very much like young human children at
play during the break in school lessons, who will hop, skip and
hand-clap with their fellows, at random, for a minute or less. The
other auks, guillemots and razorbills, do not play these games.
All guillemots appear to suffer from the nervous disease chorea,
which is not surprising—living and raising chicks on the ledges
they choose must shatter the strongest of nerves. Razorbills are
quieter; they shuffle and nod a great deal but they do not, as the
guillemots, spend minutes on end trying to spew up their own
skeletons. None knows better than the puffin how to live a
relaxed life. That's a bird without a worry in the world, happy to
the end of its life, even if it does finish in the stomach of a greater
black-backed gull.

The Lunga puffins differ in one respect from those on St. Kilda
and the Shiants. At Treshnish there is no evening fly-past. The
puffins about my camp went early to bed. Long before sunset
they had flown in and were soon underground. Rarely was a

puffin seen after sunset. I could hear them lowing, contentedly I supposed, in their burrows. They were most secretive about entering and leaving, would never do so if aware that I had an eye or lens on them. I tried rigging a camera with remote control, spent precious hours patiently waiting, and not a picture did I get. Using a hide I might have been successful. The failure was all the more exasperating because when the birds were in the open I could watch and film as I pleased.

Gulls excepted, the birds were surprisingly tame, and unless I made a sudden movement, or was noisy, they rarely flew away. Propriety demanded that a certain minimum distance be maintained, usually 7 to 10 feet. Anything less agitated the puffins into flight. Razorbills, guillemots and kittiwakes were as tolerant, but then it was impossible to approach them any closer without tumbling off a cliff, when watcher and watched would have both been in flight. One shag was particularly stoic; he wasn't leaving his nest, not for all the fish in the Sea of the Hebrides. He even submitted to the indignity of having his head patted, just sat there as though glued to his nest. Behind him were a couple of youngsters in down, writhing skinny necks, quivering bills, squeaking shrilly; "go on dad, make him go away dad, what is he dad? I don't like him dad, I don't want my photo taken dad, he was here yesterday dad, frightened mum he did dad. Pretend you're a fulmar dad and spit at him; I don't like the smell of him dad, he isn't fishy."

I placed the camera as near as I could, 32 inches, minimum focus distance with a three-times telephoto lens, and filmed the shag gaping, writhing, darting his head. Camera looking down his yellow gullet. Big close-up shots, full frame. On the screen, larger than life.

The main concentration of Lunga's birds is on the western side, grouped about the Harp Rock, a stack reaching, probably, 200 feet or more and separated from the island by about 25 feet at the nearest point. This narrow chasm is Kittiwake Alley, for that elegant cliff hanger, the kittiwake, has colonized the sheer faces of the stack and the opposite cliffs. I counted 1,000 birds and could see at least 300 nests. There were probably more on the seaward side of the stack and on cliff sides that were out of sight; possibly between 500 and 750 pairs nested here. In greater numbers were guillemots, puffins and shags; razorbills were not plentiful, although there may have been many more on the talus that lies on the shoulders of the stack.

The Harp Rock is not flat topped. It is a huge, five-sided pillar of solid rock near enough to 300 feet wide and 150 feet deep. The seaward vertical face is a great deal higher than the island side so that the top slopes steeply—in other words a cylindraceous irregular pentahedron with the top sliced off at an angle. (The bod who taught me geometry would have been thrilled to bits at the first part of that description, and thrown a fit of the shakes at the remainder.) However, because it is what it is, the stack is called the Harp Rock. Perhaps I can find a photograph and then you will know what it is you are meant to know. Enough! The birds it is that matter.

Kittiwakes breed only on the Harp Rock and the opposite cliffs, where they occupy the small niches and narrow ledges on the bare rock face, laying two eggs in a seaweed nest poised in space.

The cliffs chosen by kittiwakes as nesting sites are sunless, or so placed that the sun reaches them for only short periods of time; a sensible precaution as incubating birds, and later the chicks, are protected from long periods of summer heat. The choice makes difficulties for the photographer, especially when using colour stock. Compensating colour filters are not the easy remedy; considerable experience in their use is needed if the results are to be satisfactory, and the increased exposure time required can be a serious disadvantage. A very fast emulsion such as high speed Ektachome, used with electronic flash, is an alternative. The results can be disappointing. Neither filters nor flash help the cinematographer, he is left truly groping in the dark. The use of reflecting materials to direct light on to the subject is not often practicable, particularly for the lone worker. Winds render them useless. When working on a cliff colony a camera-man needs a dozen pairs of hands. Half gales are not uncommon, and I have often worked in a gale-force wind— that's 40 m.p.h., with gusts coming in at considerably higher speeds. In such conditions the tripod has to be forcibly held down, and camera shake is inevitable. Filming on an open cliff in a high wind can be exciting, the camera man is hanging on by his eyebrows. Common criticism of films about remote islands and the sea-bird colonies is that they are unexciting, they lack climaxes, the presentation is even. The excitement lies in the making of the film, not the viewing.

Easily the most numerous of birds at the Harp Rock, almost certainly on all Lunga, are the guillemots. On the rock itself, on and amongst the detritus piled high on left and right shoulders, are hundreds. On the wide ledges of the cliffs below the top are great huddles, in dozens and scores, crammed to capacity, squirming, writhing, flying off, flying on. Somehow they manage to incubate an egg and brood a chick. All too often their efforts are wasted. Guillemots are careless breeders, seemingly quite indifferent to the fate of eggs which go spinning off the ledge on to the rocks below, or into the sea. Careless nursemaids too. I stood at one cliff edge, above a ledge, watching a solitary guillemot and its day-old chick—the egg tooth was still visible. The ledge was not flat and level but had a nasty seaward slope and an uneven surface. The parent stood unconcerned about its offspring a foot or so away. The chick made a move towards the adult and lost its balance when trying to surmount a small step on the rock surface, fell on to its back, feet and legs bicycling madly. The adult turned and looked, and looked away unconcerned. The chick's kicking sent it rolling down the ledge and away it slid over the edge into the sea. The adult looked again, paused a second, let out a tremendous, heart-rending squawk and dived headlong to the sea.

Puffins and razorbills came in with loads of small fish dangling from their beaks. Not so the guillemots—their catch was much larger single fishes, nipped crossways between the mandibles. I saw one come in to the edge of a scrum with its catch and there it stood for minutes on end without moving. Suddenly a deft twiddle, much too quick to see how it was done, and the fish moved through 90 degrees and partly down the bird's throat. Another long pause of minutes, then another sudden gulp but still a quarter of the fish protruding from the bird's beak. And so it stood. Five minutes ticked away, six, seven, eight, nine, and the guillemot stood unmoving. Dead, I thought, choked itself by sheer gluttony. In the tenth minute the bird came back to life, gulped hard and swallowed the rest of the fish; a further period of rest, feeling, perhaps, like we do after too generous a Christmas dinner. Finally, minutes later, the guillemot moved again, bent down to touch a chick gently on the head as if to say "Yes dear, Mummy's swallowed the lovely fish; be patient, I'll cough your share up later."

A large number of puffins nest on the rock, living in burrows in the tussocky roughland across the centre. Sentinel above stand greater black-backed gulls, mortal enemies of the poor puffin.

The rock was an exasperating attraction, at its nearest point about 25 feet away, quite inaccessible without special bridging facilities. Once, 35 years ago or more, some adventurous yachtsmen tried to get across. Alistair Gibson told me the story. They bridged the gap with the mast of the yacht then went across on it, a crazy, acrobatic stunt. When they were halfway over the mast snapped.

Shags by the score nest around the cliff edge, some on ledges below the rim, others amongst rocks and boulders further back. The entire colony resented my presence, their barking outrivalling Battersea Dog's Home. One bird, a small creature and the most talkative of the lot, had a croupy voice and hoarsely uttered his piece. Not before had I noticed how greatly adult shags differ in size. Small birds, although breeders, do not appear to have attained maximum size; the differences between small and large shags is considerable. One nest contained seven eggs. The *Handbook* allows three normally, sometimes two, even six. Two females have been known to lay in the same nest. Had some ancient mariner of a shag brought home a mistress from distant waters?

Reputed to be breeding on the Treshnish Isles are petrels and shearwaters. On Lunga I found the barest of evidence that petrels, probably stormies, were present, but not the vaguest clue that shearwaters might be. Night is the time for a shearwater and petrel hunt—dry, moonless nights.

The Cruachan is a likely breeding site for both species, although there is not sufficient accommodation for a large colony. A few petrels might nest in the stones of the abandoned cottages in the villages. I found none. They and shearwaters could nest in the detritus on the shoulders of the Harp Rock. I did find a solitary shearwater in such a location on Dùn of St. Kilda.

I made one night excursion to Cruachan, and only one. Rain

stopped a planned second; and sheer, physical exhaustion caused others to be abandoned. After fourteen hours' safari by day I was more than ready to sleep come midnight, and was reluctant to lay up by day to be walking fit by nightfall. Later, I realized what I should have done. The time will come to put it into practice.

The one excursion made was not very rewarding. The night was a dull one. A bone-chilling wind, from which there was no retreat, came off the sea and blew across the hillside; there are few places on these open cliffs where you can shelter from wind and rain. After a long, tedious wait there was some bird activity, very slight. Silent forms, dark and small, flitted about the hill-side, some distant, some around my head—twisting, twirling, typical small petrel flight. Not a cry was heard, nothing but the wind and the distant noise of the sea washing the rocks below. There were long waits between bouts of activity. Stormies, I decided. The wind, the tedium, a hint of rain and heavy eyelids drove me tentwards. The only other evidence of stormies were petrel-like noises heard on the stony beach of Lunga's north-western shore.

For days I waited patiently for the wind to drop to a whisper so that I could take a tape recorder to the Harp Rock and record the auks and kittiwakes. Eventually there came an evening of summery peace and I sat upon the puffin-strewn cliff heads, the recorder on my knees, taking down the murmurings of the guillemots, razorbills and puffins. The kittiwakes remained silent. I "kittiwarked" myself hoarse in vocal encouragement and the birds sat and looked at me with that kind and gentle expression that is so characteristic of this gull. Not a single utterance in response. So I jumped up and danced for them, might as well go completely off my chump. The unexpected movement sent a huddle of several dozen guillemots into the air, which in turn launched a couple of shag flying boats, and they, flying danger-ously low, loosed the tongue of a nesting kittiwake into a sharp reprimand. Chain reaction. In a moment the recording tape was whizzing from reel to reel, taking down the most lively of cliff colony concerts blaring out at *fortissimo fortissimo*.

Golden moment. I rewound the tape, pressed the replay button. Black despair. The recording was completely ruined by a loud, background hum. Gloomily, I sat on the clifftop amongst the shags and opened up the guts of the recorder. Nothing apparently wrong, but the hum remained. Later—too late—the fault was discovered. The recorder uses a rechargeable battery or will run off the mains. In the field, when the rechargeable battery is flat, dry batteries are substituted. I had fitted a new set whilst on Lunga, and this is where the fault lay, a simple fault that could have been corrected in seconds—just that one battery was making poor electrical contact with a spring strip, hence noise, reproduced on the recording as a hum, Something to be remembered for evermore.

Late on the Saturday evening at the end of my first week, I received one of the biggest surprises of my life. There I was, alone on a desert island, looking out to sea from a cliff top and munching a chunk of Dundee cake, when, from behind me, a voice bid me good evening. Puffins may be sea parrots, nevertheless I have never heard one talk. So! I had seen this man before, three days ago. I had been over to the village searching for the well and had watched a small boat, a dinghy with an outboard motor, creep along the north-western beach. The occupant stepped ashore, pulled up his boat, waved me a far-off greeting, and busied himself carrying goods to some spot or other. A fisherman, I decided; attending to lobster pots, maybe. Later he left, in the boat, and was lost to view. Here he was again; no fisherman, a visitor, like me. Curiously enough we had not met, although we were both roaming the island and he had paid a previous visit to my camp, which he had found by chance during his wanderings and when I was absent.

We fell to talking. He was camping down on the north-western shore and was a regular visitor to Lunga, usually in October, seal time. Birds he said he knew little about, but he revealed a surprising knowledge of them. After the opening civilities I put my leading question. Did he know the locality of the well? He did; and led me to it, a place I would not have found in a lifetime and not so far from the village.

The spring lies under the shelter of an overhanging rock, very secluded, on a hillside. I drank three mugs full, one after the other. Delicious, cool, sparkling; at that moment surpassing the finest of oak-shoot wine. I filled my water container and left it on the rock, because I was walking on to my new companion's camp, and would retrieve it on my return. I was twenty minutes locating the place again.

Joe is my companion's name. Joe Reid, a Scot, from Renfrew. As he talked he spoke of boats and canoes, and I felt instinctively that our paths had crossed. Was he on Shiant last year, in June? He was indeed. Yes, they did get to Stornoway, safe and sound —a sundrenched trip. He showed me his camp, and his boat, self-built in three weeks, just the job for visiting the other islands. He had boated over from Mull, alone. We would meet again, go to Fladda next Wednesday, my last complete day at Treshnish.

We left at noon and spent the afternoon walking round Fladda. The approach to the landing beach is up a long creek; we carried and dragged the boat the last 400 yards because the tide had ebbed and the water had run out.

Fladda is the second largest of the Treshnish Isles—flat-topped, cliffs up to about 80 feet high, and almost completely surrounded by a rock pavement. In the autumn the seals come in to breed, using the flat rocks as their nursery.

Fladda is almost two islands separated by a low causeway, extending through, northwards, to a lagoon, where we found families of eiders swimming contentedly in placid water. On either side rise low cliffs, the eastern and western parts of the island being isolated one from the other by the causeway. We made a circuit of both sides. There is nothing of great interest. Some flowers but less in variety and quantity than those of Lunga. The birds are few in number and specie. A small number of shags were breeding, a herring gull colony of about thirty pairs occupied one corner, a few greater and lesser black-backed gulls elsewhere. A hoodie flapped off one headland, two lapwings circled another. Wheatears and pipits darted from rock to rock. In the sea, black guillemots; one or two scuttled off the rock pavement. They probably breed on Fladda. We searched one or two likely places, heard strange squeaky noises from beneath boulder piles, but saw no birds. That was Fladda; not nearly so interesting as Lunga. Curiously enough, Fladda was not so parched as Lunga; the rock pools were brimming.

We had no opportunity to land on Cairn na Burgh More and its sister Cairn na Burgh Beg nearly a mile away; time and tide were against us. We had to remain contented with the view through binoculars. Gulls were present but no other birds were seen. Burgh More has a ruined castle, Beg a fort. We left Fladda on a rising tide and looked wistfully at the Burghs; a choppy sea decided us that we would be sensible to head for Lunga. The little boat sat low in the water, she rode well; frequently splashy waves leapt over her bow and inboard.

"Hope you can swim," said Joe. I shook my head.

About us were a score or so of black guillemots, swimming together in packs of five or six. Halfway to Lunga we met terns fishing, and a cloud of them hung over an islet, probably their breeding territory. I couldn't determine whether they were arctic or common terns; the differences are small. We hoped we might go ashore on their islet, Sgeir an Fheoir, and take a closer look, but there was no place to tie the boat so we sheered off and instead landed on Sgeir an Eirionnaich. Herring gulls own this bit of rock and resented our visit. From the Sgeir we had a wonderful view of Lunga's northern shores and the western cliffs, saw how Cruachan spreads its hump from west to east and the terraces descend in almost geometrical perfection. Only a Pythagoras would find fault. Away in the distance is another hump, the speck of Dutchman's Cap.

The seas were building up and running fast, crashing over the rocks of Eirionnaich, urging us to leave. Reluctantly we did so. The sun shone brightly and warmed us; this was an evening of rare perfection. As we gained the shelter of Lunga the seas subsided. We nosed in. On the shore was a great circle of oyster-catchers, a score or more, heads inward, a trilling choir. So often have I heard them, distantly. This was my first close sight and it lasted but seconds; our coming disturbed the birds. Away they went, a cloud of black and white minstrels of the shore.

We pulled the boat on to the beach. I said goodbye to Joe— next morning I was to leave—climbed the lower slopes of Cruachan, walked for the last time past the Harp Rock, lingering a while to watch the kittiwakes, puffins, guillemots, razorbills and shags. The orchis were fading, the bluebells drooping. Slowly, reluctantly, I tramped across the moorland to my side of the island; the black back raised its voice, the oystercatcher left its nest—there were two eggs now, a gull or a hoodie stole one on the Sunday—and down the bank I went. My camp site was in deep shadow, the breeze was chill. I shivered. There was packing to be done. Some of my goods could be taken over to the landing place and the exertion would warm me. Next morning I must be abroad early to break camp.

The sun rose above Mull's peaks, painted Lunga pale gold, warmed my tent and brought me to a drowsy wakefulness. Time I was abroad. There was packing to be done, camp to break. For the last time I looked at the oystercatcher's nest. The previous evening both eggs were pipped; this morning the chicks' beaks were visible at the egg hole. A pity to be leaving—they would probably be free this day.

There followed a couple of hours of tedium—packing, carting goods over to the landing place. Then, nothing to do but sit, or stand and gaze around. Alistair Gibson was to retrieve me at eleven; without my watch I had no idea of time. No boat in sight. Was there time to photograph the 'catcher's nest? There wasn't. Round the headland, unexpectedly, came a boat, distantly yet and moving fast. A bigger boat than I expected; not the little craft that landed me, this was a 30 to 50-ton fishing boat. She came in, stood off about a hundred yards, loosed her dinghy. Several people scrambled in. I picked out Alistair Gibson at the oars. She was *St. Ebba*, owned by Bruce Howard, from Tarbert, Loch Fyne. He and friends had been to St. Kilda. *St. Ebba* skippered by Alistair Gibson, now they were on the homeward run to Fionnphort, picking me up on the final leg. The boat party set off to walk across Lunga. *St. Ebba* would pick them off the northwestern beach. My bits of goods went down into the dinghy. Two journeys were needed. Joe walked over to see me off. We said our goodbyes again. I jumped down into the dinghy, went aboard *St. Ebba*. We took the walking party off the beach, turned and set course for Fionnphort. By two we were standing off; at four I left for Craignure and the ferry. Before 6.30 I was at Oban.

In the morning the long trek home began, in easy stages. Unexpected adventures lay ahead. The car, a hired one, was a tired and troublesome beast, and died on me 80 miles from home, finishing the journey at a tow rope end. At 2.15 a.m. I slipped into bed. Less than six hours later I was on a London-bound train. That's the way to finish a holiday—no break, straight from Hebridean turmoils to the peace and quiet of a London office. No more raucous shriek and scream of wild sea fowl; only the gentle, soothing tinkle of telephone bells.

Twelve months to wait.

SEVEN

A Summer Shieling

··◈··

Tanera Mhor this place is called, the largest of the score or more Summer Isles that lie in the mouth of a great seaway, Loch Broom, in Wester Ross.

Once there was a village scattered across the hills and grassy terraces, a schoolhouse and a pub. Once there was a prosperous fishing station; it fell into disuse a century ago when the herring left these waters.

Now it is a deserted island, a fragment of ancient rock; not bare but heather clad, bracken too, and flower strewn; a place where sheep do graze, a haunt for birds. Their cries, the baaing of the sheep, the surge of the sea upon the rocks, these are the only voices to be heard as you sit idly by the shore in the noonday glare. At night come others, a-weening in the wind.

Down by the Anchorage the owner has built for himself a summer house, with all mod cons: pumped in and purified water from a lochan; an electrical generator, diesel driven, to operate the pump and light the house; bottled gas for cooking.

I have my humble tent, and permission to use the forsaken schoolhouse upon the hill.

Long before I landed upon Tanera an inward feeling of disappointment clutched at me. I know this emotion, have sensed it before. Nothing tangible, an indefinable something which dampens the spirit of high adventure; a strange persuasion, not ominous, not warning, not frightening, but possessive; ill brooding but not ill boding.

There is nothing forbidding about Tanera. None of the aloofness with which St. Kilda greets its visitors, no unfriendly stare

(*Above*) Hebridean wren taking food to its chicks. The nest can be seen below a large stone, part of the wall of a ruined cottage. (*Below*) Summer Isles. A buzzard's eyrie on the north-west cliffs of Tanera Mhor

Summer Isles. (*Above*) Weathering of the rock at the southern end of Tanera. (*Below*) Burnet rose holding its own amongst deep heather and bracken near the Bay of Thorns, Tanera

from dizzy heights, frowning down from the smouldering, smoky grey of a sunless dawn. Over Tanera hangs a brooding sensation of calamity, not of impending doom. Death has come and gone; the pall remains, heavily draped across the hills and moors. Somewhere in the hollows lurks a restless spirit awaiting exorcism; not a ghost of the eye but one that haunts the mind.

As I stepped on to the rocks of the northern arm of the bay, called the Anchorage, and over to a green terrace, sheep scattered wildly, bleating plaintively. In a patch of bracken lay one of their kind, a heap of fleece and bare bone. Oystercatchers swept down on me screeching, "Go away, go away". As I trudged up the hill to the schoolhouse I half expected to meet, descending, a knight at arms "alone and palely loitering . . . so haggard and so woebegone", and on the ridge above, la belle dame herself, long-haired and wild eyed. (Was she the first hippy?)

Tanera Mhor has been sold (wrote Ian McLeod, boatman at Achiltibuie, in reply to my letter, asking could he put me ashore). The owner does not like campers on the island, but the factor might give you permission to land. If not I will put you on Tanera Beag which has lots of birds and water.

The factor said come, you can use the old schoolhouse for shelter. And here I was trudging up the steep hill to inspect the place. A bare and austere building, beginning to wilt before storm weather. The window glass had long since gone. All but two of the windows were wood boarded, the two were open to the winds. I tried a door—locked and barred. There was another—I couldn't open it; I was half glad, the house didn't appeal to me. Back at the landing place I collected my goods, piled them on a close-cropped, green terrace and erected my tent. A lovely spot, made for a lone camper, surrounded by the sea and a rocky shore.

I was to return to the schoolhouse, driven by necessity. Life in a small tent, shared with such vulnerable equipment as cameras and tape recorder, was impossible. Outside the tent there was no shelter. I had brought a camp bed but not a Li-Lo and spent my first night sleeping on the ground because the camp bed was too large for the tent. Next morning I trudged once more up School-house Hill, and this time was successful in opening a door—its temperament needed understanding. Inside was desolation. I

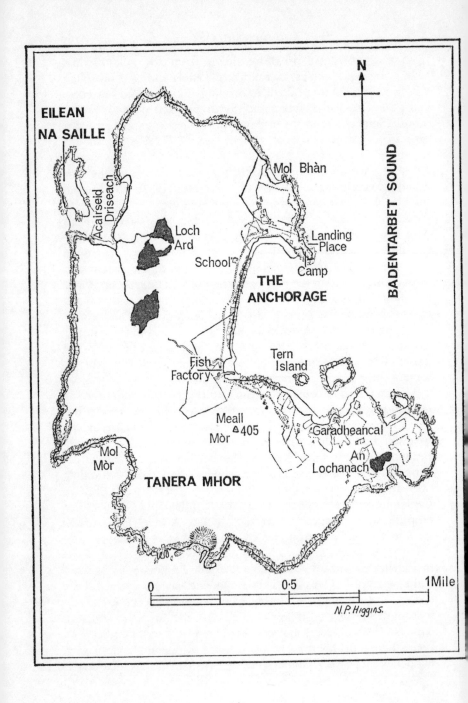

N

EILEAN
NA SAILLE

BADENTARBET SOUND

Mol Bhàn

Acairseid
Driseach

Loch
Ard

Landing
Place

School

Camp

THE
ANCHORAGE

Tern
Island

Fish
Factory

Meall
Mòr △405

Garadheancal

An
Lochanach

Mol
Mòr

TANERA MHOR

0   0·5   1Mile

N.P. Higgins.

cleaned the place up and decided I would move in. There was plenty of room. I could spread myself. The windows were glass-less, open to the weather. My bed I put into another room, perpetually in the dark because the window frames were boarded. After one night I moved the bed elsewhere, the gloom over-powered me. I needed light. In this ugly, empty house the dark, broody feeling gripped me more tightly, more strongly than it did in the old manse on Hirta. I moved my goods up, those I needed, and kept the tent as a food storehouse, so that I need not slog up the hill n-times, merely fetch such groceries that were needed. I stayed in the schoolhouse for the duration of my visit, slept well and undisturbed but never settled down comfortably, leaving each morning with a sense of relief and escape, returning each evening with reluctance.

This house has stood a century, although it has not been used as a school for more than half that time. There has been an occasional tenant of temporary residence, but not for many years. In the past two decades the place has given shelter to island visitors for a few days or weeks. They have treated the building roughly and callously. Not all of them are unknown. Dozens have left their names and townships, scribbled on the walls, carved upon the woodwork, proud of the destruction they have wrought. Instead of offering thanks for good storm shelter they have responded savagely, rending the guts out of the place —sheer and wanton destruction. Wall panelling, cupboards, doors, stair banisters, floorboards, all have been torn and wrenched away—burned, no doubt, to provide a warming fire. Man is the only animal that destroys his own habitat. Very soon this house will crumble. Rot has set in, walls reek damp, the roof leaks; storms wet and dry, allied with time and human destructiveness, have begun the demolition. Unless it is rescued, only a few years are left to it. The twin chimney pots stand awry, the cement rendering flakes away from the outer walls, the wind whistles through its poor old bones and chants a dirge.

Equalling the uninspired architecture of the house is its setting; combined, they are the very essence of bleakness. The house stands alone on a hill, 150 feet above the sea; a rough pasturage lies before it and to the north, outbuildings are at the rear. The soil is thin and black; an encroaching moor of stunted heather

surrounds the southern and western flanks. The saving grace is
the view to the east and south. In the immediate foreground is the
Anchorage. Two miles away is the mainland, the houses and
cots of Achiltibuie straggling along the shore, lacking any beauty.
For that quality you look beyond the crofters' fields and meadows
to the mountains of Sutherland and Ross. They stand in groups,
distant and lofty, of incomparable wild beauty. In writing of the
English downland the late H. J. Massingham sings the praises
of the chalk hills, to the detriment of those of stone, which, he
suggests, are "rugged, broken, abrupt and shapeless". There is
nothing shapeless about Cul Mor, Stac Polly, Canisp, Suilven
and Quinag. Here stone has been moulded into a line of beauty
that would be difficult to match.

I looked upon them in all lights, saw them in every mood.
When a watery dawn turned night into grey day they were
distant, pale ghosts. When the morning sun poured liquid gold
into Badentarbet Sound they stood out clear but mysterious. In
the burning sun of a hot midsummer afternoon they shimmered,
satin silver. Clear days brought them nearer; no longer were they
shadowy giants but clean, clear cut and bold, sometimes menacing, at others beckoning invitingly. On a storm day they hid
amongst the clouds, raised grey veils of mist over their heads.
On these days the world disappeared; there were no mountains,
no shore, no land, no sea. Excepting the bank on which I sat,
no world of any kind until the sun, fighting for supremacy,
aided by its fierce ally, the wind, brought back the mountains—
tops detached, unreal, unrecognizable. Then the low-lying land
cleared and above it sailed huge barges of grey and billowing
clouds—mountain slopes without tops, mountain tops without
slopes, sunshine above, shadow below, all in wide-screen super-
panoramic vision and glorious natural colour, browns and greens,
yellows and reds, blues and purples that no painter could ever
mix, no photographic emulsion could ever register. And when
that vision faded from sight, the day was done, night had come.
Tomorrow would paint the canvas anew.

Scenic beauty does not all lie distantly. Tanera has its share.
The loveliness of this island is mixed; raw, fierce, gentle, soft,
wild, all at once, singly and jumbled, but never any artificial
sophistication. The mask is ever changing, at one moment

attracting, the next repelling; a distant view, unbelievingly en-
chanting disintegrating into common earthiness when approached. The island is a self-contained Highlands of Scotland
in miniature; there is hill upon hill, fold after fold so that you
are for ever ascending and descending, no respite for aching leg
muscles. Great tracts are heather covered, much of it stumpy
growth of only a few inches, elsewhere reaching nearly to shoulder
height. Where the heather loses its hold bracken encroaches.
There is no great acreage of pasture, no rich soil, mostly a thin
layer of peat; not an island of fruitfulness. The spread of heather
is tremendous and covers many acres of the island in a dark
mantle. On the unchoked open spaces many species of wild
flowers grow, not in the splendid profusion of display that
Lunga of Treshnish knows but spread unostentatiously. Here a
small patch of purple orchis, on the damp bank of a wee burn a
bed of forget-me-not, in the lee of a collapsed cottage a tall clump
of foxgloves; cotton grass waving in bog, solitary dandelions,
bunches of clover. The most exciting floral discovery was burnet
rose, a score of small bushes which had defeated the heather and
sheltered from the sea behind a shambles of rock.

More or less in the centre of the northern part of Tanera are
three lochans, two together, like Siamese twins, one apart, and
a couple of burns running away to a sheltered bay, Acairsed
Driseach. The schoolhouse lays but 300 yards from Loch Ard,
the twin lochans. From these I fetched my water because the
well in the old village to the north of me was dry. The journey
was not a pleasant saunter, but a rough cross-country stumble,
over moor, uphill, downhill, a roundabout path and nothing of
great interest to distract attention from the dull monotony of
heather and stone. The lochs are shallow; they harbour giant
tadpoles, three or four times the size of those I gathered from
streams as a schoolboy. Apart from these monsters, and an
occasional gull sailing, the lochs are almost lifeless.

Tanera's real charm is discovered along its coast, and for an
island of only 800 acres the coast line is unexpectedly long. With
some licence the island may be described as kidney shaped; there
is a bight, the Anchorage, on the eastern side. Nowhere are the
cliffs lofty; they rarely climb above 100 feet. They are not thronging with seabirds but are silent and empty. Indeed, excepting

arctic terns, there are no seabirds, the true pelagics that is, on Tanera.

There is an exquisite loveliness in the rock of the cliffs, for in exposed places they have weathered into unexpectedly beautiful shapes and formations. Some are ribbed horizontally, others carved and sculptured, hollowed and honeycombed, with very thin walls, thinner and more fragile than fine porcelain ware. I was fascinated by the intricate shapes, and half afraid to lay hands on the rock face as I scrambled over the cliffs lest I despoil Nature's handiwork. That such cunning workmanship is entirely due to haphazard and freak action of weathering is unbelievable.

A great part of Tanera's shores is indented with small bays and tiny coves. At the bays bracken reaches out to the edge of gently shelving and curved beaches of stone, coarse sand and mud. The coves are enclosed by cliffs, some sheer, some stepped, some angled. Small trees, commonly rowan and birch, grow on the slopes, sheltering clumps of bluebells and primroses and adding a woodland atmosphere.

To the west of Tanera lie half a dozen sgeirs and seven islets, all part of the Summer Isles group. The largest is Tanera Beag, 210 acres; the smallest, Eilean Choinaid, is less than 6 acres. Eilean a' Char is furthest west, about a mile distant; Eilean na Saille is only 100 yards away. The remainder of the Summer Isles lie farther off, to the north, west and south.

Na Saille is the only island accessible from Tanera Mhor without a boat. At low tide a rocky, mud flat is exposed, a meadow of seaweeds across which you can walk. As this flat is left uncovered for several hours, the sea floods in slowly, and na Saille is but 12 acres in extent, there is ample time for exploration.

A stony, rocky beach rises from the flats, Caolas Eilean na Saille, and gives way to a pleasant green terrace of grass along two sides of the island. The western and northerly shores are reefed, edged with low cliffs, scarcely reaching to 20 feet at the highest place. The centre of the islet rises to a low hump, very little higher than the cliffs, but is nothing more than a forest of bracken. A few sheep wander over at ebb tide to clip the sparse pasture, staying overnight if the tide cuts off the return path.

Bird life is scant; a few scavenging gulls, a couple of pairs of oystercatchers patrolling the rocks, arctic terns zooming from end to end. During my visit a pair of ringed plovers entertained me for a long time, leading me to believe that they had eggs or young nearby. When I tired of searching and moved off they followed me, and fooled me again. At the other end of the islet, whilst I was pottering about the rocks, some aerial missile whizzed over my head so closely, so unexpectedly that I was momentarily startled. A sharp war cry, "kee-yaah kee-yaah", identified the attacker, an arctic tern. Courageous birds; in attacking an intruder they are even more persistent than herring gulls and great skuas. Under attack I quartered the ground for a nest or young running free and found nothing. Not until I left the vicinity did the tern leave me alone.

Before I left the Summer Isles I saw them all, from a boat close in, and landed on Tanera Beag for the all too short time of two hours.

Weather permitting, Ian McLeod runs cruises around the islands several times a week. He embarks his passengers at Badentarbet pier on the mainland, near the beginning of Achiltibuie's straggling hamlet. I asked him to call at Tanera and pick me up. On the appointed day I sat patiently watching a lively sea smashing at the rocks, and binoculars focused on the distant pier. McLeod set off but the course he set was not for Tanera; I leapt to my feet, waving arms and a handkerchief, executing a don't-forget-me-dance, and hollered myself hoarse. At last the boat turned to Tanera and not without difficulty managed to avoid the surges of the sea and nose in close enough for me to scramble aboard. McLeod admitted he had forgotten me, but never revealed whether he suddenly remembered or if my antics on the shore drew his attention. The trip was a delightful and bewildering cruise amongst a maze of small islands and skerries, some low-lying, others with sheer cliffs, not outstandingly high but very impressive when seen close in from a small boat. I tried hard to identify the islands from a map, and was hopelessly lost. There is not a proliferation of birds about them. Small colonies of shags and fulmars breed upon the cliff ledges, nooks and cavey holes, and gulls nest on the grassy headlands. Terns dived and soared over the small, flat islets, oystercatchers piped and probed. In the sea black guillemots, red-throated divers and eiders sailed and fished. On one skerry, in company with two shags, was a juvenile cormorant, resplendent in white waistcoat.

Tanera Beag thrilled me. I half wished I had chosen it for my summer station. There are serene bays, delightful green terraces, sheltered by cliffs of a low sort—elsewhere these cliffs shelter fulmar and shag colonies. Gulls nest where you would expect to find them, amongst the outcrops of rock and the cushions of pink thrift. Wheatears and pipits constantly flit over the heathland, search the rocks, hawk flies and scold interfering visitors. Two hours was all too short a time.

Because the seas were unfavourable I was unable to re-land on Tanera at the point where I had pitched my tent. Instead I went ashore on the far side of the Anchorage, down where the owner had built his summer house. The walk back to the schoolhouse was lengthy, winding and rough, and the last 400 or 500 yards up the steep, heather-covered hill, exhausting. For once I was glad to see the schoolhouse. Within minutes a pot of Handa Stew was gently simmering, filling the musty, decaying schoolroom with a rich and spicy aroma. A mug of oak-shoot wine invigorated tired limbs and a wind-hammered brain. A Chopin nocturne, from the desert island disc collection, echoed through the hollow, empty house, stilling moody spirits. I was at peace with the world.

Re-energized I went out to do battle with the herring gulls. They were being worked on.

I had already explored that corner of Tanera by the southern side of the Anchorage, when I had gone down to pay my respects and give my thanks to the factor and his wife—Geoffrey and Mary. They welcomed me royally and told me a great many interesting facts about Tanera, which they had learned during their eighteen-months residence. Mary had a book about Tanera, *Island Farm*, written by Dr. Fraser Darling, and describing his experiences of living and farming on the island over a period of several years. I declined Mary's kind offer to lend me the book because I had not the time to spare for reading. There were many, many other calls upon my time in that short space of two weeks.

After I had returned home I borrowed the book from a library. It has long been out of print. Certain passages left me astonished, notably Fraser Darling's reaction to the schoolhouse and the feeling of gloom he experienced at certain places on the island. My feelings and reactions echoed his. The great curiosity was that two entirely different people who are unknown to each other should have the same reactions about a particular place— and not at about the same period of time but separated by a long interval of thirty years.

Discussing this strangeness Fraser Darling wrote

> At the back of everything I get a sense of great age, dark things done, and secrets held. This sense of the feel of places is queer. It hardly bears description in a gathering of intelligent people, but it cannot be disregarded in the Highlands. Explanation should be sought along physical and physiological lines, but the psychological result is nevertheless there to be contended with, and cannot be escaped.

I have pondered over this many an hour, seeking an explanation, never finding a complete one. Somewhere in my mind is a partial answer, floating, fulmar-like, in windless air. I dislike ugliness, dark, forbidding scenes, the man-made squalor in industrial centres, the drabness of our modern planned cities and towns, twentieth-century citadels of grime, spreading, far too rapidly, to spoil the natural loveliness of country villages. The despoliation offends the eye of some and depresses the spirit. And yet many people are unaffected by the sight, even when living amidst it.

On Tanera the dark moorland, the dreary architecture of the schoolhouse, the utter desolation of the part-ruined cots were a shock to my inward senses as well as to my eye. Amongst them I was depressed; away from them there was no gloom. Fraser Darling truly says "Once rot has begun in a place, human beings soon take on the habits of carrion crows and are not satisfied until they have pulled it to pieces."

Why, when I—the physical I—am not part of these places, and stand on the edge, aloof, why am I so passionately affected within by what I see. That is the part that lies outside comprehension.

> I love waves, and winds, and storms,
> Everything almost
> Which is Nature's, and may be
> Untainted by man's misery.

That's what is wrong with these parts of Tanera, they are tainted by man's misery.

The factor's house was a pleasant abode, much more so, I thought, than the owner's shieling. His place lay down by the sea, on a low coastal flat, protected from behind by a green knoll. Geoffrey and Mary lived at the opposite side of the bay, a cosy, sheltered nook on a ledge on a hill. Small, enclosed gardens lay at the back and front of the house and were planted with shrubs and flowers. Spacious windows looked over the Anchorage, and a tiny conservatory added some protection against turbulent wind. The living room and conservatory were gaily decked with cultivated flowering plants, adding a touch of the exotic, so unexpected in this wild place, and contrasting strangely with the island's natural flora of bog, moor and rock.

In the Anchorage are two small islands, Eilean Mhor, scarcely 5 acres, and Eilean Beag, probably not even an acre. A colony of about forty pairs of Arctic terns possessed Beag; Eilean Mhor supported a few gulls, including a greater black back. Geoffrey told me that he had watched a greylag raise her family there during the past spring. One by one the goslings disappeared, most likely into the black back's stomach, and when the goose left in May she took with her one gosling, the only survivor.

Ducks probably nest on this island—I never got on to it or the smaller one—for eiders and mergansers were to be seen fishing regularly in the waters of the Anchorage and further out in Badentarbert Sound. I found only one eider's nest, deep in heather, with five eggs. That was elsewhere, on a low, rocky islet beyond the owner's house, and which was isolated from Tanera at high tide. This was a pleasant place to laze upon because it commanded superb views of distant mountains, the north-eastern part of Tanera and many of the other Summer Isles to the south. In addition to the eider, a few pairs of herring gulls were nesting. The eider was deeply sunk in the heather, only her head floating above. She was a nervous bird, and would flush when I was 20 or 30 feet away. To get a close-up view I had to watch her through binoculars and move slowly, otherwise she was off the nest in a flash, leaving her eggs at the mercy of the herring gulls. She looked a young bird, probably was, and inexperienced. When flushed she flew out into the bay, swam round a rocky point and, when there were no intruders, quickly returned to the nest. My experience of other eiders—limited to a few birds—is that they will only leave the nest when taken by complete surprise, and will return very warily, slowly, in stages.

Although this was the only evidence I found of eiders breeding upon Tanera I often saw them swimming offshore—but none with chicks. They seemed to be late breeders, if at all. Most likely they were late. When I first arrived on Tanera I was surprised to see eider ducks *and drakes*, on the rocks and swimming together. This must have been on the eve of the departure of the drakes because latterly none was seen. In several years wandering in the Hebrides, always in the latter half of June, I have never before seen an eider drake at this time.

Ornithologically Tanera was disappointing—meaning that I missed the unruly mob of seafowl, the primitive cliff dwellers, too idle, too ignorant to build a nest and safeguard the egg of the year. Before I went to the island I knew that old friends were not resident. Time, I told myself, to go looking for birds, not to sit goggle-eyed on a cliff, for ever watching auks steam-rollering their way across ledges. Tanera had birds in variety, but not in quantity.

I would go a day and see nothing more than a few scolding wheatears, as many pipits and a dozen gulls. Other days were enlivened by unexpected sightings. Once, as I sat on a grassy bank below the schoolhouse, scanning the bay for ducks and divers, a cock hen harrier flew across and alighted on the rocks near my tent, staying just long enough for me to identify it; then it flew off, a bird of passage. Another one-day-only sighting was an immature common gull, flying up and down inshore by the southern cliffs of Tanera. I mentioned that to Geoffrey; he had never seen one about at any time. Once, some thirty years ago, a colony nested on one of the islands in the Anchorage, so Fraser Darling recorded. He also said that the arctic terns colonized the second, smaller island, where they still breed.

In the roof of a derelict cottage, standing between the owner's and the factor's houses, two pairs of swallows had built their nest, feather-lined with the mud still damp. In other nooks were abandoned nests, blackbirds perhaps or a song thrush. Once I

thought I saw a thrush near the schoolhouse, just a few seconds'
glimpse as it rested momentarily on a fence wire, not long enough
to identify the bird exactly. Possibly it was a redwing, could have
been a missel thrush.

One late afternoon there was a bit of a storm. All day there
was a threat, thunder rumbling distantly Ullapool way, slowly
moving northward, circling to the west beyond Tanera, heading
south, then returning north from the west and reaching Tanera
at about tea-time—by town and city standard, that is. A lazy
storm that rumbled slowly around the island, as though on an
errand of duty. Thor, probably irritable, was stamping his feet,
tossing out an occasional bolt just to establish who is boss. There
wasn't much rain. There had been none, of course, for the
customary six weeks. Fields, meadows and crops were parched.
Rain was needed. Next year *they*—Authority, who handles these
matters—would have the water pipe laid, then there would be no
reliance on rain. Next year? Might be the one after, or even two.
Who can tell? *They* move slowly. There was wind, blowing
noisily—at low force, not up to Hebridean standards. Above
thunder crash and wind howl rose another voice, sweet and clear;
not continuously but in sudden bursts, followed by long silences.
The music puzzled me, so I recorded it. As yet I have not been
able to identify the singer. I went out into the open searching,
and not a bird was to be seen; wherever it was perched, it remained well hidden. I can only guess at the identity. Who else
besides a storm-cock sings in storm?

There was a family of hooded crows to be seen daily—two
adults, two juveniles—flying, perching, resting, feeding. Hoodies
are crafty birds, and very difficult to approach. Sometimes they
will permit you to draw reasonably near, and then, when you
try to outwit them by clapping a 400-mm lens on the camera,
you find yourself outdone. At the moment the lens is focused
sharply, the hoodie utters a derisive croak and is airborne.

One afternoon I found their nest, by then abandoned. The
structure is a large mass of twigs, cosily lined with sheep's wool.
This one was built on the ground, on a cliff top, at the very
edge, hidden by coarse, tall-growing grass, and protected by a
slender monolith of rock. This same cliff gives shelter to the
burnet roses I had found and is on the western side of Tanera,

separated by a placid stretch of water, Acairsed Driseach, from
the islet of Eilean na Saille. The English of Acairsed Driseach is
the Bay or Anchorage of Thorns. The only thorny plant I found
about here was the rose, no hawthorn, no blackthorn, no other
thorn of any sort. Bracken and heather are in control.

Gaelic names, to me, have a musical lilt and a romantic quality.
The English of them can be image destroying. Often I sat or
stood on Schoolhouse Hill, gaze wandering idly over distant,
hazy mountains, and coming back again and again to the Anchorage, enjoying the peacefulness, drinking in the beauty.
From my lofty, distant perch the scene was idyllic: deep, blue
waters, soft green pastures dark stained by heather patches;
everything toylike, two small, white painted cottages nestling
amongst hillocks, diminutive grey sheep dotting the green, two
boats rocking gently in the lazy lapping of waters; all shimmering
in midsummer sunshine of late evening; that green place down
there, the Gaels call Garadheancal. How lovely to be able to tell
your friends, I have a house at Garadheancal. How soul destroying
when you learn that this heaven-sounding word means the
Cabbage Patch.

South-west of this earthy spot is a lochan, An Lochanach, on
a coastal flat, surrounded on three sides by rising ground. At the
front a narrow, stone beach, which breaks the line of cliffs,
separates the lochan from the sea. Two pairs of ringed plovers
called plaintively, unceasingly as they flitted about the shallows;
and after a long and diligent search I found a juvenile perched
on a rock at the water's edge, still and silent, merging perfectly
into the surroundings. Had the bird not made a couple of slight
bobs of its head I might never have discovered it.

A few terns came to fish the waters. Graceful birds, quick of
eye, fast on the wing, swoops and darts following in quick
succession. They well earn their sobriquet of Sea Swallow. They
swept over and across the lochan, gliding low, a quick bank, a
sudden upward surge and then a sharp, swift dive, often to
catch, and equally often to miss, a fishy meal for hungry chicks.

Swimming serenely in the centre of the lochan were eight
young shelducks, not long out of the egg. Expert divers and
swimmers at that early age. When I arrived the parents flew
anxiously about, endlessly circling, dropping on to the beach for

seconds only and then away on another patrol. The chicks played happily, unmindful of danger, of greater black back and herring gulls lurking in the shadow of rocks, and all-destroying man plodding clumsily by the water's edge.

On a later visit the shelducks had put to sea, and I counted only four ducklings. The duck was with her family, the drake was on flying patrol. In from the sea, down one side of the narrow inlet, across the lochan, circle it, out to sea following the opposite side of the inlet; back to the family, flying close to the water, calling softly at regular intervals—a low, two syllable note half warning, half reassuring. To and fro; I counted a dozen circuits and then he splashed down near the family. Not for long; he was off again. Despite this constant watch his family diminished. On my last visit, two days before I left, there were only two ducklings. If only they would stay together more would survive. When danger threatens, they dive and they scatter. On surfacing they are yards away from the protecting adults. Predatory gull has only to hover nearby, and swoop when an isolated duckling surfaces.

Climbing up from An Lochanach and over a series of low hillocks, which are the green, round tops of coastal cliff, I disturbed a nesting oystercatcher, sitting on two eggs. The bird flew straight ahead, uttered no cry. Unusual to meet a silent oystercatcher; mostly they are screeching their heads off, attacking valiantly. When I had thought about it I recollected that noisy oystercatchers are protecting territory, squawking alarms to chicks and juveniles. The noise also appears to be made to intimidate trespassers. The bird dives at you, utters its war cry and passes by, often falling silent until it turns, then cries and attacks again. Silent departures from the nest may be a protective measure; noise would attract attention and hence reveal the location of the nest to predators. Gulls and hoodies relish an egg.

Descending slowly and silently down a hillside to a small beach, I surprised a large, brown bird. Instantly it was gone, silently. I had never been so close to a buzzard before. This one had very pale plumage. There was no likelihood of a nest being found here, the cliffs were low and not wide ledged, there was no cover of protecting vegetation. Probably a hunting buzzard, taking a rest. In *Island Farm* Fraser Darling wrote "An increase in the population of buzzards in this area is greatly to be desired. Unfortunately the ravens have bullied the buzzards so much that they have not nested on Tanera since 1940."

I saw ravens flying, usually high, sometimes low, to the annoyance of the hoodies. Never did I see a raven alight. They may not come to Tanera any more. One evening I saw a flight of sixteen large, black-plumaged birds, flying very high and south. Too high to be able to identify them definitely as ravens. I dismissed the thought that they could be crows. There is an old saying, common in Southern England, that if you see a lone rook, that's a crow; and if you see a lot of crows together, they are rooks—another way of emphasizing that crows are lone birds and rooks are sociable, and thus a ready means of identification. Excepting when crows are gathered round an animal's carcase in feast, I have never seen more than two adult crows together. When I mentioned the flight of sixteen to Geoffrey he thought it probable that they could be crows, as in the north of Britain if nowhere else they do move about in flocks. I must ask our computer, the uncontradictable know-all.

Two days after I had seen the pale buzzard, I came face to face with another, a dark one. I was climbing round a headland, about 20 feet below the rim, following a sheep path through the heather which grows very dense about here, and a little above where the cliff falls sheer away to the sea almost 100 feet below. This place is on the northern end of the island at Creag Ard.

The bird lifted easily, climbed swiftly and high, and began to glide and soar around the headland, mewing plaintively, "peeioo, peeioo, peeioo," with just a short pause between each cry. I watched it through the glasses for several minutes and all the time it mewed. Then there came another "peeioo", distantly, and down from the north; flying high came a second buzzard, the mate. They came together, crying continuously, soaring and gliding, for half an hour, and then one bird flew off; the other stayed on patrol. In the meantime I decided that they had an eyrie on this headland, and I was going to find it before I left. An hour later I began to wish I hadn't been so cocky. I searched that headland inch-by-inch and realized, belatedly, that blindly hunting, hoping for the best, was useless.

The headland is horseshoe shaped and bisected by a narrow

wall of rock. Below is stone beach, in which heather and bracken grows near the cliffs and small trees grow on the cliff sides. The crown of the headland is covered with heather, growing to a height of more than 4 feet—terrible stuff to walk through. I found a way down to the beach—quite an exciting one—and began to scan the cliff sides through the glasses, looking for movement which might indicate young buzzards. Suddenly, great excitement, I could see something moving, but couldn't detect what. By climbing up the bisecting wall I could find out. I went up, not without difficulty, which brought out a little perspiration on the brow, not entirely due to exertion. The climb was fruitless; just vegetation moving in the wind.

Back again to the beach. At last it dawned on me that the sensible and obvious action was to look for whitewash on the rocks and vegetation. Within a few minutes the glasses picked up a patch at the top of the far side of the cliff. Back to the top of the headland, laborious but not difficult. Now to get round to the whitewash patch. Laborious and difficult. Two hours after surprising the buzzard I had found the eyrie. All I had to do was to struggle back through the heather to the other side of the headland, retrieve the cameras, and return.

The eyrie was on a ledge on the cliff face, about 75 feet sheer above the ground, and 30 feet below the cliff top. The approach was one-way only, the last several feet overgrown with heather, 3 or 4 feet high and growing to the cliff edge. I cleared a way through to the brink of the cliff, the heather folding back easily and forming a springy mattress, on which I could lie and look into the eyrie. There sat two chicks, well feathered, three perhaps four weeks old. Fierce-looking little birds, eyes bright and stony. I expected them to show anger; they just ignored me. One sat facing me, immobile, unblinking, passive. The other was a couple of feet away, equally aloof.

The nest was not what I expected to find, not a great structure of sticks, not much more than a bed of dried vegetable matter, covering the ledge where the chicks lazed and where lay remnants of food. What they were being fed on I couldn't make out, all I could see was a small pile of red and raw flesh. The ledge went back into the cliff, cavelike, providing a shady retreat when the sun was hot.

I put out a hand to the nearest chick, couldn't reach it by 18 inches, and the chick ignored me. To test its fear or anger reactions, I pushed one tripod leg within an inch or so of its bill; no response of any sort. I touched the bill, and was snubbed. So I gently pushed the tripod leg beneath the bird and against its breast. This elicited some response, the chick shuffled uneasily and gave the tripod leg a reproving nudge with its bill, gently, not a vicious snap, and sat and stared at me, unconcerned. Obviously I wouldn't get action photographs from this pair. The results were disappointing, dead and dull. Simply a record—buzzard chicks in eyrie. I thought they might be overfed and sleepy. On a later visit they were just as immobile, in the cave, larder empty.

The photographs might have been better had I been able to choose a better viewpoint, got a frontal look. There was no possibility of erecting a hide and hoping to record the parents visiting. There was barely room for me, tripod and camera. Two tripod legs could be placed on firm rock, the third had to be wedged in a springy, heather mattress. By planting right foot firmly on solid rock, right hand grasping more rock, left foot over the cliff edge supported by a knobbly protuberance, and lying on the heather mattress I was just able to peer into the camera viewfinder and use the left hand to operate the shutter. Most exhausting, very exasperating. I can't remember ever having taken so much trouble to get such indifferent photographs and film shots. Buzzards, huh! if only the little wretches had screamed at me I would have been gratified. The species has a reputation for sluggardly behaviour.

The herring gulls were the birds that provided action and excitement, ferociously dive-bombing me when I blundered into their territories. They were not present in large numbers. A few pairs on the east coast, above the Anchorage; a few more on the south-western shores; a few isolationists. Not a full-scale gullery anywhere.

Between the northern extremity of the ruined village and the east coast, immediately above the Anchorage, was a small lapwing colony, ten pairs perhaps. There the ground is flat but rough, pasture of sorts; a terrace of grass, with bracken encroaching in large and small patches. I always saw lapwings about the place,

12

feeding and idling, and when I ventured on to their territory they
would take to the air, circling and crying for evermore. Occasionally they would mob a gull. I searched the area constantly but
never found a nest, an egg or chick. The only evidence I found
was slightly off their territory, a dead chick lying in the grass,
untouched. Probably wandered and died of exposure in night
rain; not more than 48 hours out of the egg.

The northern side of the Anchorage was colonized by several
pairs of oystercatchers. Two had nested in the ruins of old
buildings, only the footings remained. Both nests contained an
egg, just one each. One egg disappeared overnight. I was never
certain whether it had been stolen or hatched. I found no chicks.
The other egg was destroyed by a gull or hoodie; I found the
egg, outside the nest, neatly holed and empty.

A permanent resident on the coastal flat of the northern shore
of the Anchorage by my landing place, was a motherless lamb.
That heap of bones and fleece lying in the bracken 100 yards
away could have been the ewe. As is common on Scottish islands,
the sheep were shepherdless, put down to graze and grow,
representing in the owner's eyes not animals but so many pounds,
shillings and pence. The flock didn't look a wealthy lot. Too little
grazing, perhaps, for too many sheep, and Nature was taking a
toll. I have never seen so many dead sheep on an island before. A
few animals will always die from sickness, a few more from falls.
None of these had died from falls, they lay in open spaces. In a
rough shack I found a recumbent lamb, which struggled to its
feet. For a moment I thought it had a broken hind leg. The poor
creature bounded out of the shack and ran. A horrible sight. No
broken leg; the entire hindquarters were paralysed and were
dragged along the ground as the lamb galloped away. The speed
it made, although so badly crippled, was incredible. It reached a
beach, was unable to move on the pebbles, and sank down. Out
there the hoodies and gulls would have the eyes out long before
it died—a cruel death. I put the animal out of misery.

The motherless lamb never left the grazing by my landing
place; some invisible rope tethered it to the locality. Nor did the
lamb mix with the other sheep when they passed that way—
which they did regularly as they roamed around the island. It
kept aloof, never following to pastures new when the flock

moved on. At first the lamb was afraid of me and scampered off; gradually it became accustomed to my visits and I could approach within yards without causing it to run away. At night it would lie in a hollow, out of the wind, and sleep until morning light awakened it, completely alone.

I was puzzled. Sheep are beyond my ken; never before have I seen lamb or adult remain aloof from the flock. An ethologist could make something of such behaviour, perhaps; I could only guess and wonder.

Another puzzle was cuckoos. I saw three when I was in Achiltibuie, flying, calling, perching, most times of the day; they never visited Tanera just 2 miles across the sound.

There were human visitors. On a lovely evening a yacht came in under sail, as graceful and beautiful as a swan, and anchored off the small island where the terns nest. Hours later I met the crew on the top of Schoolhouse Hill, a couple of men and their wives. They had only put in for a night's stay. As we chatted I told them of a curious coincidence, that Mary had once lived in the town where I now lived. One of the women asked me where that was. I was born in the hospital there, she told me, and lived at Sidcup, a few miles further off. They sailed just after dawn.

Another boat came in from Ullapool. A motor vessel, carrying holiday makers, out for a day's cruise amongst the Summer Isles, with a landing on Tanera for two or three hours. Their boat tied up at the jetty, in the calm waters of the Anchorage, comings and goings ordered by the tides. The jetty was originally built for the fish factory, now in ruins, mere walls and a gaunt ghost of a once prosperous industry, brought to naught by the changing habits of fish. Herring, which frequented these waters, moved away and took Tanera's prosperity with them.

There is an airy walk from the jetty, up the hill, over the top, and down the other side to pass by the pleasant green spot where the factor's house stands. Tireless folk can walk up a steep slope to the 400-foot summit of Meall Mhor, Tanera's beacon, and from there look out to all the Summer Isles, down Loch Broom to Ullapool, and on a clear day see the lonely Shiant Isles, away and away to Lewis and Harris.

Amongst the old fish factory buildings, whilst sheltering from a long and heavy rain shower, I found wrens, but no nests.

They were breeding somewhere because they had beakloads of
scrumptious grubs. I met only one wren elsewhere and that I
heard, never saw. Whilst I was setting up a camera at the buzzard's eyrie there was a sudden burst of wren's song, just the
one, loud, jubilant and triumphant sounding.

I never heard a wren sing at dawn; this was because there
were no wrens at the schoolhouse and only once did I make the
effort to lop out of bed at such an hour. The choice of the day
was a bad one. There was heavy cloud over and below the
mountain tops; night was reluctant to leave. I took an alarm
clock with me for this very purpose and made a deliberate dawn
rising once only. On the chosen morning the sun was late
showing—well above the hills before a golden beam pierced
Badentarbert Sound. A bitterly cold wind came up from the
south-east and almost sent me back to bed. Real determination
was needed to face the coming of that new day. The only entertainment was a dawn chorus—no trio! I counted only three
voices, and their owners seemed to be reluctant to be abroad.

A sunset was more rewarding. Sunsets were rarely seen on any
of my island visits. There is always a great lump of the world
standing in the way. In Shetland Ronas Hill blotted out the view.
On Handa we had to climb to the top, and as often as not clouds
blotted out the sky. Soay was the obstacle at St. Kilda, the hills
of Harris at Shiant. Lunga sunsets were visible when I walked
over to the Harp Rock on a cloudless evening. On Tanera a
tedious walk up and down the heather-clad hills brought me to
the western side. On evenings of high cloud I could watch the
sun go down behind the mainland hills. Rarely have I seen the
sun go down below the rim of the sea; land masses, all too often,
clutter up the horizon. I did see a superb sunset on Tanera, the
sun finally disappearing below the hills at 10.35 p.m. I filmed the
last stages on black and white film, for no better purpose than
curiosity. When projected the setting sun is seen to be elliptical
in shape.

My last complete day on Tanera was unexciting. The diary
summation is: rose 8.40 a.m., sunny and cloudy all day, very
slight rain in the wind, warm in the sun, bed 10.30 p.m., went
nowhere, did nothing.

Did nothing! How untrue. I spent most of the time packing

and carting a pile of goods down to my tent by the landing place. Quite a trail it was up and down Schoolhouse Hill. I rarely took the most direct route but wandered afield letting minor occurrences distract me from the job in hand—the journeyings could have been done in half the time. I was sidetracked by grub-carrying wheatears, as devilish as trows. I went searching for lapwing's nests and chicks, tried to make real friends with the lamb, watched the hoodies and high-flying buzzards through the glasses. Down by the water's edge, on a wee bit bog, I found a few purple orchis. They had to be photographed.

The tent, I decided, ought to come down whilst it was dry—tomorrow could be wet. It was. I stacked tent and packages amongst the rocks, well protected by black polythene sheeting.

The evening I spent at the factor's house, saying goodbye to Geoffrey. Mary had gone on an overland safari to Inverness a day or so earlier. At about ten I wandered back my way, pausing for a while by the jetty and the old fish factory. The evening was a lovely one, still and warm; the water calm, scarcely a ripple. The midges were a nuisance—curious, because midges rarely bother me, I have always been able to defy them. On Lunga, one tribe got the better of me, drove me to distraction, not by biting but by a queer tickling sensation. Their compatriots on Tanera did the same. Always packed in the first-aid kit is insect repellent, never used until on Lunga, a very effective ointment in a restricted way. The directions for use warn that the ointment must not be allowed to come into contact with eyes. I smeared it as near as I could but these midges were a daring sort, like children who play last across the road in front of oncoming traffic. The midges sat, walked, nibbled and crowded all along the edge of the ointment, pranced over my eyelids and generally made life unpleasant. They also got into my hair. I wished I had long, cow-like ears to flick them away. Out of the sun they are less obnoxious so the long plod up the hillside and through the stunted heather was bearable.

The sun was below the horizon when I reached the schoolhouse. I made it an early-to-bed night. I had to be early-to-rise, McLeod was taking me off the rocks in less than twelve hours.

About midnight a wind rattled the schoolhouse and brought a gentle patter of rain. In two hours a full gale was ripping in from the south-east, rampaging across Tanera, wind howling, rain lashing. A steady drip spattered down from the broken roof. All night the storm clattered, easing at about six. I was about by then, finishing my packing. When I began the several treks to the landing place, the skies were grey, the rain had ceased and the wind was thumping in from the south-east, putting the sea into a great turmoil, piling it up on to the rocks by the landing place. Not much chance of boarding a boat from there. Nor down in the bay. The seas were just as lively.

There was no sign of the boat at ten, when I should have been retrieved. Nor at eleven, nor twelve. The rain came back, in light showers, in heavy downpours, in sudden squalls, and then set in for a heavy, long spell. I guessed that near 12.30 the wind might drop and the skies lighten, as this was about full tide. And so they did. The clouds lifted, there was hardly any wind. There was, however, a deep swell. I waited hopefully, in vain. An hour later the wind was back, shifting around to the south-west. More rain came, thinly, delugingly—not that that mattered, I was well protected. The tedium was the hateful part, wondering when the boat would come, whether it would come. I had no means of knowing.

Time was heavy. So was the rain, bucketing down, wind-driven; easing to almost nothing, then another deluge; dropping straight down when the wind hesitated, horizontal sheets when the wind tore back in another rip and roar outburst.

I wandered up and down, like a caged animal—around the point where my tent had been, along the rocks by the bay, across the point, over to the landing place, scrabbling over the rocks. Back again to where lay my pile of goods, dry under the polythene sheeting. Round and round, again and again, clockwise, anti-clockwise to relieve the tedium. Sometimes, for a diversion, walking further up the bay. Always walking with an eye on Badentarbert Sound, hoping to see a boat. I knew that I would not and yet dare not go far afield.

My surroundings palled, so I walked further along the coast, 300 or 400 yards, and found shelter amongst the rocks, out of the wind. There was a small beach here. I watched the tide slowly cover it. A merganser flew off the rocks onto the sea, gradually drifted away from the shore and when he reached a certain point took to the air, circled, came back to the rocks, flew onto the sea again, drifted out, just so far, flew back to the rocks and started again. Time after time. I ventured along to where he was landing. Perhaps there was a nest and a mate. I knew there was not. I had searched this length of coast for hours during the past twelve days. Herring gulls lived here. Nevertheless, I might have missed a nest, and this was something to do. There was not a nest, of course.

The merganser went away. So did I. Those lapwings, they must nest somewhere. I tramped. I searched. I scoured the bracken and the ruined cottages, both empty of any breeding evidence. I walked up the hillside to a semi-ruined cottage, the roof was part intact and gave some shelter. I stood in the doorway and scanned the pastures stretching down to the sea, swept it time and again with the binoculars. Watched the lapwings, standing, pottering, running. No sign of chicks—they must breed somewhere, they never leave the place except to fly about and around it. Wherever they nested was a well-kept secret.

Back to my shelter amongst the rocks. A large bird flew by, heading north, following the shore. I had only a rear view. Mighty like a curlew's rump, I told myself. Hours later the bird flew back—heading south, following the shore, as purposefully as it had headed north—and confirmed that it was a curlew.

The tide began to fall. The wind roared on. Rain fell monotonously. I was hungry, nibbled at chocolate, chewed raisins. A good standby, raisins. I eat two or three pounds whilst on an island.

I fell to musing. What a day to leave! The first time I had not got away on schedule. Not an exciting island, Tanera. Pleasant enough. I missed the auks. The terns were a compensation. The buzzard's eyrie was an exciting find. Not *the* most exciting. That had turned up, out of the blue, on my way to Achiltibuie, about 5 miles before journey's end.

I was motoring slowly, saw a bird, carrying food in its bill, alight on overhead telephone wire—a strange bird, to me. I stopped to watch. Another bird joined the first, the plumage distinctly different. The two appeared a little agitated, flew off, perched nearby, flew back. I moved further down the road, and watched. The first bird flew down to the bank alongside the road, disappeared into the low growing furze, emerged a few seconds later and flew off. The second bird flew down to the same place on the bank, stayed a few seconds and flew away. Two minutes or so and they were back. I got the glasses on them. The first bird had a black head and mantle, that was the feature that had first attracted my attention. The breast was orange, extending to the shoulder, the rump was white. The second bird had a browny head, orange breast and white rump. Nothing less than a pair of bramblings. Breeding too. I watched for half an hour during which time they repeatedly flew in with food. I gauged the spot they were visiting, grabbed a camera, and when the cock bird flew off, made my way to what I supposed was a nest. What I did not see was the other bird, the hen, fly in. I parted the furze at what I believed to be the nest site and all but put my hand on the hen. She was ramming food into a youngster's gape. For two, perhaps three seconds, we looked at each other, and then she was gone.

Quickly, I took two photographs of the nestlings—too quickly, the results could have been better. I had the 135-mm lens on the camera, this was a job for the 50-mm. I should have changed and duplicated the shots but did not want to cause too much inter-ference. From the car I watched the birds continuing their feeding visits. Both were quite unperturbed by my visit.

Bramblings are rare breeders in Britain. Only two records have been recognized. One, in Sutherland in 1920, when a bird was watched building in a Scots pine, 25 feet above ground. Seven eggs were laid, the last on 31st May. Another record was at Monar in Easter Ross and was identified by nest and eggs only; the birds were not seen. Other reports of breeding have come from Inverness and Perthshire. My sighting was of both parents feeding the young, the hen at the nest, and four chicks, about two days old, in the nest.

These birds normally build their nest 5 feet or more above ground, in birches and conifers. This nest was on the ground, alongside a road, in the bank, hidden by furze. The hillside here is treeless. The nest was constructed from grass and bits of twiggy material, probably heather, and lined with feathers and hair. All four chicks were blind, and covered with a light-coloured down. Disappointing that I could not substantiate the find, for official recognition. One of the penalties of travelling solo.

Geoffrey coming past in his boat, about two, interrupted my musings. He turned inshore, came as close as he dared and said he would ask McLeod if and when he was coming for me.

Two hours later Geoffrey returned, stopped and shouted that McLeod would take me off at about five, weather permitting. Another dreary wait.

At about 5.30 the boat came in. Quite a game we had embarking me and my pile of goods. The wind blew hard, the seas smashed in, trying to pile the boat onto the rocks, then drawing it back as a wave subsided. The rocks were slippery and added a touch of circus excitement as I stumbled over them, loaded with packages. We managed without mishap, and at six I was aboard. We made for Old Dornie, where I had left my car, about 4 miles away. A wet journey, rain and windlashed. Then a remarkable thing happened. As we cruised into Old Dornie both wind dropped and rain stopped, almost instantaneously. In the time I unloaded my goods onto the jetty the skies cleared, the sun shone with summer brilliance and warmth. The evening was perfect.

I loaded the car. The engine refused to start; the battery was down and the car was bogged in mud. Two hours later I got clear and motored back to Achiltibuie, there to spend the night. A piping-hot bath wrung the rain out of me. Sandwiches and cider reawakened the inner man. What a day it had been.

Next morning was wet. I stopped 5 miles up the road to look at the bramblings. There was no nest, only a gaunt, black and scorched hillside. A heath fire had destroyed what, fourteen days ago, had been a pleasant green place, flower-strewn, a haunt of bramblings. There were no corpses, no parents—perhaps the chicks had grown sufficiently to escape.

I motored on, picked up petrol at Ullapool, paused at Dundonnell, again at lovely Gruinard Bay, and looked out to the distant Summer Isles. On to Gairloch for lunch. The skies cleared, the rain ceased. I spent an hour at Loch Maree, turned south-west at Kinlochewe to motor through the Torridon mountains. The sun was shining, adding a touch of lustre, making the drive a memorable one. The Torridons are one of Scotland's most beautiful gems.

That night I spent at the Torridonian Hotel, dined sumptuously on Duck l'Orange; the memory lingers.

Next morning there was more rain. At Loch Carron I decided to turn north-east to Achnasheen, hoping to escape the rain. There was no escape. Through to Inverness, south-east to Carrbridge. A detour to Loch Garten to peep at the ospreys. Across the Cairngorms, over the Grampians. Rain all the time, clouds low, blotting out the distance. Beautiful scenery but not on show this day. A disappointing day, ending at Amulree.

Homing is a sad business. For a couple of days I linger in the Highlands. When I leave them and cross the Forth, and Edinburgh is only miles away, I know the holiday is done. The big wrench comes when I cross the Border. There stands a sign, pleading "Will you no come back again?"

Oh yes! I'll come back.

There is an island called Mingulay, at the end of the Long Island chain. I must motor to Oban, take the steamer to Castlebay on Barra; a fisherman shall land me on the rocks of Mingulay. There is a ruined village, a pile, a burn, a beach of silver sand, a green terrace where I can pitch my tent. Her hills rise above 800 feet. Off-shore stand two great stacks, 400 feet high, the ledges thronged with all my friends, the wild sea fowl.

Aye, I'll be back, come Chune.

APPENDIX A

Further Reading

Many, many books have been written about the Hebrides. Very few discuss intimately the tiny, remote isles, those specks in the seas that are outside the sphere of human affairs—lonely places affording a temporary anchorage, grazing for a few sheep, breeding haunts for wild birds. The Shetland Isles have been sadly neglected. This list of recommended books, is therefore, a selected one. A great deal of information has been published in scientific papers, to which the general public will not have access; purposely these have been ignored.

Shetland:
> *The Shetland Isles*, by A. T. Cluness (Robert Hale, London 1951). The author is a Shetland man; the book is all embracing.
> *The Far Flung Isles*, by Garry Hogg (Robert Hale, London 1961). A personal account dealing with the present-day affairs of Orkney and Shetland. The latter is dealt with separately in the second part of the book.

> *Orkney and Shetland*, by Eric Linklater (Robert Hale, London 1965). Complementary to the above-mentioned books and filling in the gaps. A new edition (1971) brings the story right up to date.

St. Kilda:
> *St. Kilda Summer*, by Kenneth Williamson and J. Morton Boyd (Oliver and Boyd, Edinburgh and London 1960). This is the classic book on St. Kilda. Tells of the coming of the Armed Forces into the fastness of the islands, and continues with lively accounts of St. Kilda's history, its geography, archaeology and birds. To know St. Kilda you must live there. Read this book before you go, read it again when you return, read it

once a month for evermore, then you will be as enthusiastic about these lovely islands as the authors and the members of that unique association, the St. Kilda Club—membership is confined to people who have lived on the isles for not less than twenty-four hours.

The Scottish National Trust organizes party visits, the finest packaged holiday that anyone has offered. Gives you everything at a cost of about £30 and knocks all other so-called holidays for the adventurous into oblivion.

Life and Death of St. Kilda, by Tom Steel (National Trust for Scotland, Edinburgh 1965). A history of the St. Kildans and their way of living since early times; ending with the evacuation in 1930.

Summer Isles:
Island Farm, by Dr. Frank Fraser Darling (Bell, London 1944). The author lived on Tanera Mhor for several years, as a farmer.

Other Islands:
Mentioned briefly and at length in other books by various authors. Another classic is *Island Going*, by Robert Atkinson (Collins, London 1949). He visited and describes Handa Island, The Shiants, St. Kilda and many other of the small islands. Don't fail to read this.

Mosaic of Islands, by Kenneth Williamson and J. Morton Boyd (Oliver and Boyd, Edinburgh and London 1963), covers several Hebridean islands and the Faroes. Two chapters and an appendix are devoted to St. Kilda.

The Highlands and Islands, by Dr. Fraser Darling and J. Morton Boyd (Collins, London 1964), is just what the title says it is. Covers, very comprehensively, almost every aspect of Scotland's natural history. Mention is made of the Shiants, Summer Isles and Treshnish Isles.

Portrait of Skye and the Outer Hebrides, by W. Douglas Simpson (Robert Hale, London 1967) is primarily concerned with Skye, touches upon some Hebridean outliers and has a chapter devoted to St. Kilda.

The *Vertebrate Fauna* series, by H. J. Harvie-Brown (Douglas, Edinburgh 1887–1904), is approaching a centenary. They make interesting reading because of the comparison of conditions existing a hundred years ago and those pertaining today. Some remarkable changes have taken place in bird populations since Harvie-Brown's expeditions.

Finally there is Martin Martin, who travelled extensively around the Hebrides and visited St. Kilda. He is the author of *A Description of the Western Isles of Scotland* (London 1703), and *A Late Voyage to St. Kilda* (London 1698). Martin is very readable. Regrettably his references to the lesser isles are all too brief.

Most of these books are out of print; you will have to borrow from and through public libraries.

APPENDIX B

BIRD LISTS

There is nothing I can add to the comprehensive lists published elsewhere for Handa, Shetland and St. Kilda. Therefore, I have omitted them.

My lists for Shiant, Summer and Treshnish Isles are not exhaustive. They are restricted to the birds that I identified during my two weeks' stay on these islands. Only those species of which the young or nests with eggs were seen, with or without adults, are classified as 'breeding'.

The flowers, animals, insects and other subjects that attracted my attention are mentioned only in the foregoing chapters. I have not attempted to compile lists. My knowledge of these subjects is limited; many were seen, few were recognized. I was not seeking them, there wasn't time. Those creatures and flowers that I did see aroused my interests, have led me to widen my knowledge of them and so increase my enjoyment. I have been forced to realize that I cannot restrict my interest to birds alone; the flora, the fauna, the rocks, all things are intricately interwoven. In short, I find it impossible to remain a simple ornithologist, circumstances have forced me to become a bewildered naturalist, of woefully limited knowledge, hung with binoculars, tape recorder and cameras, bulging with pocket handbooks that will help me to widen my appreciation of "all things bright and beautiful, all creatures great and small".

SHIANT ISLES

17th–30th June 1967

GARBH EILEAN

FULMAR *Fulmarus glacialis*
Breeding. Sitting on eggs on ledges. Plentiful.
SHAG *Phalacrocorax aristotelis*
Breeding. Main breeding colonies amongst the boulder scree

on the east shore. Smaller colonies on the west and north shores. Many hundreds. Eggs and chicks seen at all locations.

GOLDEN EAGLE *Aquila chrysaëtos*

One sighting. Brief appearance over east shore, not more than a minute. Harried by oystercatcher and greater black-backed gull. Not definitely identified. A large brown bird, larger than the black back. Might have been an immature.

OYSTERCATCHER *Haematopus ostralegus*

Breeding. A few pairs flying and feeding about the island. One pair seen with a juvenile near the sheep pen. Juvenile captured and released.

SNIPE *Gallinago gallinago*

Seen flying, disturbed whilst feeding, heard calling and drumming. A few single birds only sighted.

GREATER BLACK-BACKED GULL *Larus marinus*

Breeding, western side, fair number. Eggs and chicks found.

LESSER BLACK-BACKED GULL *Larus fuscus*

Breeding, eggs and chicks. Two pairs amongst a herring gull colony on the west slopes. Exclusive colony, about fifty pairs, on the north-west slopes.

HERRING GULL *Larus argentatus*

Breeding. Eggs and chick. Fair number in small colonies, mostly on the west shores.

RAZORBILL *Alca torda*

Breeding. Eggs and chicks. Big colony amongst the boulder scree associating with puffins and shags on east shore. Smaller colonies on north-west cliffs. Very plentiful, probably numbered in tens of thousands, and about three to four times greater than puffins.

GUILLEMOT *Uria aalge*

Breeding, eggs only seen. A few breeding on ledges on east shore, amongst the scree, associated with puffins and razorbills. Larger colonies on north shores, but these difficult to see from the island.

BLACK GUILLEMOT *Uria grylle*

An occasional bird seen in the Bay of Shiant between Garbh, an Tighe and Mhuire.

PUFFIN *Fratercula arctica*

Breeding. One chick seen just hatched, still very wet, under a boulder on the east slopes. Adults seen going to ground, with fish, amongst scree. More birds nesting under the boulders than in burrows. Main concentration on the east shore slopes. Very plentiful, estimated, very roughly, about 20,000 pairs.

ROCK DOVE *Columba livia*
One pair seen flying twice. Could have been same pair.

SKYLARK *Alauda arvensis*
Seen on the ground, and flying. Heard singing. Never more than two seen at one time. A few scattered pairs.

RAVEN *Corvus corax*
A pair flying regularly around and over the island.

WREN *Troglodytes troglodytes*
Breeding. Adults seen feeding juveniles amongst rocks on west shore near landing beach.

BLACKBIRD *Turdus merula*
Breeding. Hen adult seen with three juveniles, feeding and flying.

WHEATEAR *Oenanthe oenanthe*
Several seen, flying, flitting, hawking flies and perching.

PIPIT *Anthus*
Probably rock but not definitely identified. Several flying around the islands and cliffs.

STARLING *Sturnus vulgaris*
Breeding. Flock of about thirty birds, adults and juveniles seen flying about the rocks and ruins.

AN TIGHE

SHAG *Phalacrocorax aristotelis*
Breeding, eggs and chicks. Many pairs. Less plentiful than on Garbh. Some nests isolated.

EIDER *Somateria mollissima*
Breeding. One nest with three eggs, all hatched successfully, found amongst rocks near herring gull territory. Several adult ducks seen offshore swimming with chicks.

OYSTERCATCHER *Haematopus ostralegus*
Seen feeding, and flying over the island. Mostly feeding on west coast.

TURNSTONE *Arenaria interpres*
Small party landed once, on rocks near the cottage.

SNIPE *Gallinago gallinago*
Discovered feeding in boggy places. Single birds only seen. Occasionally heard drumming and seen flying.

GREATER BLACK-BACKED GULL *Larus marinus*
Breeding. Eggs and chicks. Main colony at southern tip of island, Mianish. Fair numbers. A few mingled with a herring gull colony west shore, mid-island.

HERRING GULL *Larus argentatus*
Breeding. Eggs and chicks. Small and large colonies all the way along the west coast. Fair numbers.

KITTIWAKES *Rissa tridactyla*
Breeding. Small colony, about fifty birds, on the east cliffs towards Mianish. Birds seen on nests. Impossible to approach closely to check for eggs and chicks. A boat needed to make a closer observation.

BLACK GUILLEMOT *Uria grylle*
Single birds. Seen near shore, east side, swimming.

ROCK DOVE *Columba livia*
One, perhaps two, pairs seen flying.

SKYLARK *Alauda arvensis*
Heard singing. Seen on the ground and flying. Few pairs only.

RAVEN *Corvus corax*
A pair occasionally circled the island.

HOODED CROW *Corvus cornix*
Four birds flew regularly about the island.

WHEATEAR *Oenanthe oenanthe*
Fairly common about the island.

PIPIT *Anthus*
See Garbh.

TRESHNISH ISLES
15th–28th June 1968

LUNGA

FULMAR *Fulmarus glacialis*
Breeding. Sitting on eggs, moderate numbers nesting all around the island, many on isolated ledges.

SHAG *Phalacrocorax aristotelis*
Breeding. Eggs, chicks, juveniles, many of the latter flying

and swimming. Very plentiful, probably numbered in thousands. Rocks, cliffs and skerries always fringed. Nests on cliff
ledges, rock outcrops and inland cliffs.

EIDER *Somateria mollissima*
Believed breeding. No nests found. Seen in the coastal waters
with chicks.

SHELDUCK *Tadorna tadorna*
Believed breeding. Two adults and four chicks seen once in
coastal waters at the north of the island.

OYSTERCATCHER *Haematopus ostralegus*
Breeding. One nest found with three eggs. One egg disap-
peared. The other two hatching on the day of departure.
Several pairs of adults seen about the island.

RINGED PLOVER *Charadrius hiaticula*
Breeding. One nest with two eggs found. Later, eggshell seen
but not chicks.

GREATER BLACK-BACKED GULL *Larus marinus*
Breeding. Eggs and chicks. In colonies and isolated nests.
Fairly plentiful.

LESSER BLACK-BACKED GULL *Larus fuscus*
Breeding. Eggs and chicks. Small number, one colony of about
twenty pairs near a herring gull colony.

HERRING GULL *Larus argentatus*
Breeding. Eggs and chicks, isolated pairs and small colonies.
Fairly plentiful.

KITTIWAKE *Rissa tridactyla*
Breeding. Adults at nests with eggs and chicks. Concentrated
at the Harp Rock area. Very plentiful.

RAZORBILL *Alca torda*
Breeding. Eggs and chicks. In small groups and fair-sized
colonies. Plentiful.

GUILLEMOT *Uria aalge*
Breeding. Eggs and chicks. Concentrated on and about the
Harp Rock. Very plentiful.

BLACK GUILLEMOT *Uria grylle*
Occasionally a few birds—up to a dozen—seen swimming in
the coastal waters.

PUFFIN *Fratercula arctica*
Breeding, all on the northern half of the island, in burrows and

amongst scree, in small groups and large colonies. Very plentiful. Main concentrations near the Harp Rock. Seen flying in with fish and taking fish to burrows. One burrow excavated, nestling, with egg tooth, found.

ROCK DOVE *Columba livia*
Single bird seen twice, flying.

SKYLARK *Alauda arvensis*
Breeding. One nest found with eggs.

RAVEN *Corvus corax*
A pair seen, occasionally, flying over and around Lunga.

HOODED CROW *Corvus cornix*
Four birds seen regularly flying about Lunga and standing on cliffs and rocks.

WREN *Troglodytes troglodytes*
Breeding. One nest found, with chicks. Juveniles seen flying.

WHEATEAR *Oenanthe oenanthe*
Fairly common. Seen flying, feeding, hawking flies and carrying food in bills.

ROCK PIPIT *Anthus spinoletta*
Breeding. One nest with young. Adults also seen entering nest—beneath a large stone in the ground—with food. Birds fairly common.

STARLING *Sturnus vulgaris*
Breeding. Flock—twenty to thirty birds—of adults and juveniles, flying over the island and roosting on the rocks.

TWITE *Carduelis flavirostris*
Occasional pair seen near my tent.

FLADDA

SHAG *Phalacrocorax aristotelis*
Breeding, eggs and young. Small number.

EIDER *Somateria mollissima*
Believed breeding. Swimming near the shore with chicks.

LAPWING *Vanellus vanellus*
Two birds seen circling.

GREATER BLACK-BACKED GULL *Larus marinus*
LESSET BLACK-BACKED GULL *Larus fuscus*
HERRING GULL *Larus argentatus*
All three breeding. Eggs and chicks. In small colonies.

BLACK GUILLEMOT *Uria grylle*
Seen amongst the rocks on the beaches. Small parties in the sea.
HOODED CROW *Corvus cornix*
One seen.
WHEATEAR *Oenanthe oenanthe*
Several birds flying about the island.
ROCK PIPIT *Anthus spinoletta*
Several birds flying about the island.

SGEIR AN EIRIONNAICH

EIDER *Somateria mollissima*
Swimming in surrounding sea.
HERRING GULL *Larus argentatus*
Breeding. Small colony with eggs and chicks.
BLACK GUILLEMOT *Uria grylle*
Swimming in surrounding sea; small parties.

SGEIR AM FHEOIR

TERN *Sterna macrura* or *hirundo*
Colony of about forty pairs. Not possible to land and check whether breeding. This appears to be the only tern territory at Treshnish. Whether arctic or common tern not ascertained. Believed to be arctic.

SUMMER ISLES

14th–25th June 1969

TANERA MHOR

EIDER *Somateria mollissima*
Breeding. One nest with five eggs found on southern part of the island, in heather. Adult ducks seen in the Anchorage bay. Drakes also seen.
RED BREASTED MERGANSER *Mercus serrator*
A few single birds swimming inshore.
SHELDUCK *Tadorna tadorna*
Breeding. One family only seen, duck and drake with eight ducklings swimming in a lochan.
BUZZARD *Buteo buteo*
Breeding. Eyrie with two well-feathered chicks on a cliff ledge, 75 feet above sea (about) in north of island.

HEN HARRIER *Circus cyaneus*
One bird, cock, flying northwards, alighted on the rocks
north side of Anchorage. Seen once only.

OYSTERCATCHER *Haematopus ostralegus*
Breeding. Nests with eggs found by Anchorage and on south-
west side of island. Several pairs seen about the shores and hills.

LAPWING *Vanellus vanellus*
Breeding. Several pairs flying on rough pasture, bracken etc.,
north of Anchorage. One dead chick found.

RINGED PLOVER *Charadrius hiaticula*
Breeding. One pair seen at a small lochan. One juvenile alone,
seen at the same lochan. Another pair seen on the islet of
Eilean na Saille.

SNIPE *Gallinago gallinago*
A few birds heard and occasionally seen. Disturbed when
feeding.

CURLEW *Numenius arquata*
One bird seen flying north along the east coast. Later, one bird,
probably the same, flew along east coast heading south.

GREATER BLACK-BACKED GULL *Larus marinus*
A few pairs seen about the Anchorage. Believed to be breeding
on the small island, Eilean Mhor, in the Anchorage.

HERRING GULL *Larus argentatus*
Several pairs breeding, widely separated usually, along east
coast of southern half of island.

COMMON GULL *Larus canus*
One juvenile seen flying on one day along the south-west coast.

ARCTIC TERN *Sterna macrura*
Breeding. Small colony, about forty pairs on the small island,
Eilean Beag, in the Anchorage. Also terns seen on islet of
Eilean na Saille.

BLACK GUILLEMOT *Uria grylle*
A few birds seen swimming offshore, particularly in Badentarbert Sound.

ROCK DOVE *Columba livia*
Occasional bird seen flying over the island.

SKYLARK *Alauda arvensis*
A few birds heard singing, and seen flying and running over
the island.

SWALLOW *Hirundo rustica*
Breeding. Newly built nests in an old building south of the Anchorage. Eggs not then laid.

RAVEN *Corvus corax*
A pair seen occasionally.

HOODED CROW *Corvus cornix*
Believed breeding. Four birds, often together, seen about the island. Disused nest found on north cliffs. Evidently had been only recently evacuated.

WREN *Troglodytes troglodytes*
Believed breeding. Two birds seen at old herring factory, carrying food in bills. One bird heard at buzzard's eyrie.

THRUSH *Turdus*
Thrush-like song heard during a thunderstorm. One bird seen once, no opportunity to positively identify. Small, might have been a redwing.

WHEATEAR *Oenanthe oenanthe*
Several pairs seen about the island.

WARBLER *Phylloscopus* —
Believed to be a willow warbler (*phylloscopus trochilus*). Seen once only amongst small trees about the factor's house. Had an eye stripe.

MEADOW PIPIT *Anthus pratensis*
Several pairs seen about the hillsides, rarely near the sea.

PIED WAGTAIL *Motacilla alba*
Occasional bird seen on the beaches and hill tops.

OTHER ISLANDS

RED THROATED DIVER *Gavia stellata*
Seen fishing and swimming in the seas about the other islands.

FULMAR *Fulmarus glacialis*
Breeding. Sitting on eggs and nests on Tanera Beag and some other islands.

CORMORANT *Phalacrocorax carbo*
Juveniles and adults seen swimming and standing on rocks.

SHAG *Phalacrocorax aristotelis*
Breeding. Seen with young in nests. Small colonies.

HERON *Ardea cinerea*
Seen swimming, fishing and standing.

MAINLAND by Achiltibuie

CUCKOO *Cuculus canorus*
  Seen and heard. Three birds perching on power lines and posts.
BRAMBLING *Fringilla montifringilla*
  Breeding. Nest with four chicks found on ground by roadside.
  Both adults seen feeding young at nest.

Index

Tormentil, 119, 137
Torridon Hotel, 185
Torridon Mountains, 185
Traigh Shourie, 64, 67, 70, 71
Treshnish Isles, 93, 129–59
Trolladale Water, 27
Trolligarts, 23
Trondra, 47, 48
Troulligarth, 23
Trout, 19, 39
Turnstone, 91
Tushkar, 72
Twaroes, Burn of, 38, 40
Twite, 89, 93

U
Ullapool, 140, 172, 179, 185
Ulsta, 25
Ulva, 131
Unst, 25–7
Ure, Villians of, 21
Uyea, 41, 42
Uyea Sound, 25

V
Vidlin Voe, 38
Village Bay, 74, 75, 77–9, 86, 88, 96, 98
Violet, 119, 131, 137
Voe, 24, 25, 27, 32, 35, 37, 41, 48

W
Wadbister Voe, 32
Wagtail, pied, 64
Walls, 23
Warwick, Alex, 75, 79, 81, 85, 98
Water Lily, 69
Waterhorse, 68
Waterston, George, 59, 60, 107
—— Irene, 60
Watt, Bruce, 74
Weir, T., 107, 124
Weisdale, 22
Western Isles, 74, 75, 77, 78, 99
West Lunna Voe, 38
Whal Firth, 26
Williamson, K., 84, 85, 93
Willock, Arthur, 55, 59, 60
—— Ruby, 55, 60
Wind Hamars, 22, 48
Windy Brae, 32
Wren, 145, 146, 179, 180
—— St. Kilda, 85, 93, 95

Y
Yell, 19, 24–6, 27, 37
Yell Sound, 38, 39

Z
Zetland, 24